# Advances in Intelligent Systems and Computing

Volume 753

**Series editor**

Janusz Kacprzyk, Polish Academy of Sciences, Warsaw, Poland
e-mail: kacprzyk@ibspan.waw.pl

The series "Advances in Intelligent Systems and Computing" contains publications on theory, applications, and design methods of Intelligent Systems and Intelligent Computing. Virtually all disciplines such as engineering, natural sciences, computer and information science, ICT, economics, business, e-commerce, environment, healthcare, life science are covered. The list of topics spans all the areas of modern intelligent systems and computing such as: computational intelligence, soft computing including neural networks, fuzzy systems, evolutionary computing and the fusion of these paradigms, social intelligence, ambient intelligence, computational neuroscience, artificial life, virtual worlds and society, cognitive science and systems, Perception and Vision, DNA and immune based systems, self-organizing and adaptive systems, e-Learning and teaching, human-centered and human-centric computing, recommender systems, intelligent control, robotics and mechatronics including human-machine teaming, knowledge-based paradigms, learning paradigms, machine ethics, intelligent data analysis, knowledge management, intelligent agents, intelligent decision making and support, intelligent network security, trust management, interactive entertainment, Web intelligence and multimedia.

The publications within "Advances in Intelligent Systems and Computing" are primarily proceedings of important conferences, symposia and congresses. They cover significant recent developments in the field, both of a foundational and applicable character. An important characteristic feature of the series is the short publication time and world-wide distribution. This permits a rapid and broad dissemination of research results.

More information about this series at http://www.springer.com/series/11156

Mamdouh Alenezi · Basit Qureshi
Editors

# 5th International Symposium on Data Mining Applications

 Springer

*Editors*
Mamdouh Alenezi
Prince Sultan University
Riyadh
Saudi Arabia

Basit Qureshi
Prince Sultan University
Riyadh
Saudi Arabia

ISSN 2194-5357          ISSN 2194-5365   (electronic)
Advances in Intelligent Systems and Computing
ISBN 978-3-319-78752-7          ISBN 978-3-319-78753-4   (eBook)
https://doi.org/10.1007/978-3-319-78753-4

Library of Congress Control Number: 2018937397

Printed on acid-free paper

This Springer imprint is published by the registered company Springer International Publishing AG
part of Springer Nature
The registered company address is: Gewerbestrasse 11, 6330 Cham, Switzerland

# Foreword

On behalf of the 5th International Symposium on Data Mining Applications (SDMA2016) Chairs and Organizers, we welcome you to this fifth edition of the SDMA2018 held at Prince Sultan University on March 21–22, 2018, Riyadh, Saudi Arabia. SDMA, in alignment with the Kingdom's Vision 2030, is organized by Prince Sultan University, every two years, in order to advance the state of the art in data mining research and its various real-world applications. The Symposium provides a forum for academia, industry and governmental organization and agencies to highlight common research problems and collaborate to provide scientific insights into possible solutions.

The Symposium provides opportunities for technical collaboration among data mining and machine learning researchers and enthusiasts within the Kingdom of Saudi Arabia, GCC countries, and the Middle East region. Since 2010, SDMA has been organized routinely, every two years, with previous symposia held in 2010, 2012, 2014, and 2016. The Symposium is gradually establishing as a venue for quality research presentations and a forum for academics, government, and industrial organizations locally and internationally. This year, SDMA2018 received 58 submissions out of which 18 were accepted for presentation and subsequent publication in the Symposium proceedings with an acceptance rate of 31%.

SDMA2018 continues the tradition of previous SDMAs that featured a number of distinguished keynote speakers. This year, Dr. Albert Bifet (Telecom ParisTech), an ACM distinguished speaker, is invited to deliver a keynote address. Mr. Caesar Lopez (Senseta), a well-known technology entrepreneur and CTO of Senseta, will also be delivering a keynote address. This SDMA will also feature tutorial sessions from industry sponsored by Oracle, Microsoft, and IBM. The focus of these tutorial sessions would be on big data analytics and data mining.

It is an honor to organize SDMA2018 that involved experts from around the world. The scientific committee comprises of 31 leading experts from various countries including, Malaysia, UAE, Saudi Arabia, UK, USA, Portugal, Australia, Italy, Spain, Turkey, Philippines, Jordan, Taiwan, China, Greece, and Poland. The authors and co-authors of the papers submitted to SDMA2018 originate from three continents (Africa, Asia, and Europe) and ten countries (Saudi Arabia, Malaysia, UK, Egypt, India, Japan, Lebanon, Libya, Algeria, and Tunisia).

A research event such as SDMA2018 can only be organized by the support and great voluntary efforts of many people and organizations. As with the previous symposia, this Symposium was sponsored by Prince Sultan University. We would like to thank the university management, in particular Dr. Ahmad Yamani, Rector, Prince Sultan University, for supporting this research event. This event is also sponsored by Saudi Investment Bank.

We would like to show our heartfelt appreciation to all members of the scientific committee and authors for their great effort in making this event a scientifically recognized International Symposium.

Many thanks for coming to Riyadh, we wish you a successful and enjoyable SDMA2018!

Mamdouh Alenezi
Basit Qureshi
SDMA2018 Chairs/Co-chairs

# Organization

## Organizing Committee

### Symposium Co-chairs

Mamdouh Alenezi
Basit Qureshi

### Media and Publicity Chairs

Yasir Javed
Khaled Al Mustafa

### Industrial Collaboration Chair

Ahmed Sameh

### Invited Talks Chairs

Anis Koubaa
Souad Larabi Marie-Sainte

### Local Organization Chairs

Jaber Jemai
Tanzila Saba

### Tutorials Chair

Mohammad Zarour

### Web Manager

Abdullah Al Tarsha

**Technical Program Committee**

| | |
|---|---|
| Adnan Khan | University Malaysia Sarawak, Malaysia |
| Ahmed Sameh | Prince Sultan University, Saudi Arabia |
| Ameur BenSefia | Higher College of Technology, United Arab Emirates |
| Armando J. Pinho | University of Aveiro, Portugal |
| Chris J. Hinde | Loughborough University, UK |
| Dat Tran | University of Canberra, Australia |
| El-Sayed M. El-Alfy | King Fahd University of Petroleum and Minerals (KFUPM), Saudi Arabia |
| Husni A. Al-Muhtaseb | King Fahd University of Petroleum and Minerals (KFUPM), Saudi Arabia |
| Izzat M. Alsmadi | University of Texas A&M San Antonio, USA |
| Keeley Crockett | Manchester Metropolitan University, UK |
| Liyakathunisa Syed | Taibah University, Saudi Arabia |
| Mario Luca Bernardi | Research Centre on Software Technology (RCOST), University of Sannio, Italy |
| Mohamed Alkanhal | Prince Sultan University, Saudi Arabia |
| Mohammed Akour | Yarmouk University, Jordan |
| Muazzam Siddiqi | King Abdulaziz University, Saudi Arabia |
| Mu-Chun Su | National Central University, Taiwan |
| Muhammad Badruddin Khan | Al-Imam Muhammad ibn Saud Islamic University, Saudi Arabia |
| Muna Al-Razgan | King Saud University, Saudi Arabia |
| Nor Shahriza Abdul Karim | Prince Sultan University, Saudi Arabia |
| Qasem Obeidat | Al-Imam Muhammad ibn Saud Islamic University, Saudi Arabia |
| Raed Seetan | Slippery Rock University, USA |
| Raihan Rasool | Victoria University, Australia |
| Rashid Mehmood | King AbdulAziz University, Saudi Arabia |
| Renann Baldovino | De La Salle University, Philippines |
| Rohana Mahmud | Universiti Malaya, Malaysia |
| Ryszard Tadeusiewicz | AGH, Poland |
| Sameera Almulla | Khalifa University of Science, Technology and Research, United Arab Emirates |
| Sebastian Ventura | University of Cordoba, Spain |
| Serkan Ayvaz | BAU, Turkey |
| Shadi Banitaan | University of Detroit Mercy, USA |
| Stathis Stamatatos | University of the Aegean, Greece |

## Additional Reviewers

Ankit Soni
Suad Alramouni
Tanzila Saba
Yasir Javed
Souad Larabi Marie-Sainte
Soumaya Chaffar
Sarab AlMuhaideb
Iman Almomani
Inayatullah Shah

## Acknowledgements

Organized by

Sponsored by

Supported by

# Keynote Speakers

# Massive Online Analytics for the Internet of Things (IoT)

Albert Bifet

**Abstract.** Big data and the Internet of Things (IoT) have the potential to fundamentally shift the way we interact with our surroundings. The challenge of deriving insights from the Internet of Things (IoT) has been recognized as one of the most exciting and key opportunities for both academia and industry. Advanced analysis of big data streams from sensors and devices is bound to become a key area of data mining research as the number of applications requiring such processing increases. Dealing with the evolution over time of such data streams, i.e., with concepts that drift or change completely, is one of the core issues in-stream mining. In this talk, I will present an overview of data stream mining, and I will introduce some popular open-source tools for data stream mining.

**Short Biography:** Albert Bifet is Associate Professor at Telecom ParisTech and Honorary Research Associate at the WEKA Machine Learning Group at the University of Waikato. Previously, he worked at Huawei Noah's Ark Lab in Hong Kong, Yahoo Labs in Barcelona, University of Waikato, and UPC BarcelonaTech. He is the author of a book on Adaptive Stream Mining and Pattern Learning and Mining from Evolving Data Streams. He is one of the leaders of MOA and Apache SAMOA software environments for implementing algorithms and running experiments for online learning from evolving data streams. He was serving as Co-Chair of the Industrial Track of IEEE MDM 2016, ECML PKDD 2015, and as Co-Chair of BigMine (2015, 2014, 2013, 2012), and ACM SAC Data Streams Track (2017, 2016, 2015, 2014, 2013, 2012).

# Object and Pattern Recognition in Aerial Images Using Convolutional Neural Networks—Examples and Applications

Caesar Lopez

**Abstract.** There are numerous applications of unmanned aerial vehicles (UAVs) in the management of smart cities assets and disaster response. A few examples include routine bridge inspections, disaster management, power line surveillance, and traffic surveying. This conference will present details on potential approaches for procedure and parameters used for the training of convolutional neural networks (CNNs) on a set of aerial images for efficient and automated object recognition. Potential application areas in the several fields are also highlighted. Working code will be presented and shared in a public GitHub repository. Starting from a convolutional neural network implemented in the "YOLO" ("You Only Look Once") platform, we will show how objects can be tracked, detected ("seen"), and classified ("comprehended") from video feeds supplied by UAVs in real time. This type of technology has the potential of increasing efficiency of drone data mining.

**Short Biography:** Winner of the MIT 35 Award, Mr. López has co-founded over a half a dozen technology companies in data analysis and data science-related technologies. He also served as General Manager of Gestión Jaibaná, an information management company from Bogotá, Colombia; and as General Manager of Enelar del Magdalena S.A. – a utilities provider that operates street lighting systems in Aracataca, Colombia. At the international level, he founded and managed TotalCom Tech in Ecuador, a technology advisory company for governments. He also co-founded PergaminoDB, a data analysis company in Las Vegas, Nevada. Mr. López was recognized with the 2012 Colombia MIT TR35 Award for his work on data sensors on UAV platforms. His team has coordinated over 8,000 UAV sensor flights for over 30 different locations in Latin America. He is also a nominee of Accenture's Award for innovation as well as a recipient of Colombia's 2010 Innova Award, recognized by presidential decree for having created "Colombia's most innovative company." He currently serves as CTO of Senseta, a big data analytics firm based in Palo Alto, California, and Bogotá, Colombia.

# Contents

**Social Networks (SocialNets)**

**Bioinformatics (Bio)**

# Machine Learning and Data Mining Applications (MLDM)

# Use of Machine Learning for Rate Adaptation in MPEG-DASH for Quality of Experience Improvement

Ibrahim Rizqallah Alzahrani$^{(\boxtimes)}$, Naeem Ramzan, Stamos Katsigiannis, and Abbes Amira

School of Engineering and Computing, University of the West of Scotland, Paisley PA1 2BE, UK
{Ibrahim.Alzahrani,Naeem.Ramzan,Stamos.Katsigiannis, Abbes.Amira}@uws.ac.uk

**Abstract.** Dynamic adaptive video streaming over HTTP (DASH) has been developed as one of the most suitable technologies for the transmission of live and on-demand audio and video content over any IP network. In this work, we propose a machine learning-based method for selecting the optimal target quality, in terms of bitrate, for video streaming through an MPEG-DASH server. The proposed method takes into consideration both the bandwidth availability and the client's buffer state, as well as the bitrate of each video segment, in order to choose the best available quality/bitrate. The primary purpose of using machine learning for the adaptation is to let clients know/learn about the environment in a supervised manner. By doing this, the efficiency of the rate adaptation can be improved, thus leading to better requests for video representations. Run-time complexity would be minimized, thus improving QoE. The experimental evaluation of the proposed approach showed that the optimal target quality could be predicted with an accuracy of 79%, demonstrating its potential.

**Keywords:** MPEG-DASH · Machine learning · Rate adaptation
QoE · Video streaming

## 1 Introduction

Video has become one of the most important media elements across the entertainment and communications industries. It is expected that in 2021, over than 3/4 of Internet data will be video [1]. As a result, extensive research is being conducted in order to better understand and sort out the complexities of time variations, unstable bandwidth, and end-to-end or startup delays during transmission.

The current video streaming market is dominated by three primary providers: Microsoft (smooth-streaming) [2], Adobe (HTTP-dynamic streaming) [3], and

© Springer International Publishing AG, part of Springer Nature 2018
M. Alenezi and B. Qureshi (Eds.): *5th International Symposium on Data Mining Applications*, pp. 3–11, 2018.
https://doi.org/10.1007/978-3-319-78753-4_1

Apple (HTTP-live streaming) [4]. Each of these companies has its own content format, encoding, and protocol that are firmly tied to that specific vendor, a fact that often leads to video content being prepared for delivery in these three different formats. Thus, many challenges are inherent in producing, encoding, and delivering video streams, but a solution must be provided for all vendors in the market because multimedia streaming is rapidly becoming the dominant media form [1]. Technology such as MPEG-DASH [5] provides a viable solution to the above issue by allowing inter-operability between various electronic devices and servers. The MPEG-DASH client accesses services from the web/media server via the framework described in [6]. At the server, encoded versions of the video are produced and then chunked into smaller segments that are all the same size. These segments are requested by the client via HTTP GET or partial GET requests, but show up as of the same video despite differing methods of transmission, ensuring an acceptable Quality-of-Experience (QoE).

In this work, we propose a machine learning-based method for selecting the optimal target quality, in terms of bitrate, for video streaming through an MPEG-DASH server. The proposed method takes into consideration both the bandwidth availability and the client's buffer state, as well as the bitrate of each video segment, in order to choose the best available quality/bitrate. Choosing the optimal target bitrate will allow the client to make better rate adaptation choices, thus enhancing QoE on the client's end.

The rest of this paper is organised in four sections. Section 2 provides an outline of the related published research, while the proposed method is described in Sect. 3. The results and discussion of the experimental evaluation are provided in Sect. 4 and conclusions are drawn in Sect. 5.

## 2   Related Work

The rate adaptation of Dynamic Adaptive video Streaming over HTTP (DASH) is usually based on the available bandwidth, the client's buffer, or some combination of the items above.

Petrangeli et al. [7] proposed a framework based on OpenFlow, that prioritises the delivery of specific video segments to avoid overfilling clients' buffers, a problem that may lead to the video freezing. They based their method on a machine learning approach, which relies on the RUSBoost algorithm (which is used for detecting whether a client is about to encounter a freeze) and fuzzy logic (which decides whether the circumstances of the prioritisation queue are proper to efficiently prioritise the segment). Petrangeli et al.'s proposal [7] would notice if a client is about to encounter a freeze, and drive the network prioritisation to overcome it. They carried out an extensive performance comparison of various video streaming scenarios with other HAS solutions (i.e. heuristics FINEAS and MSS). Their conclusion showed that the proposed method could decrease the video freezes by about 65% and freeze time by about 45%. However, their proposed method is based on the collected network measurements, and ignores both the information about the streamed videos and the clients' system characteristics.

Chan et al. [8] proposed a rate algorithm to improve the efficiency of bandwidth utilisation based on the transport-layer information, which is explored as a means of estimating the buffer size on the client's end. The proposed algorithm showed better results compared to Apple HLS [4] and Microsoft smooth streaming [2]. However, the proposed algorithm lacks the accuracy and up-to-date details of the adaptation process that are available with additional HTTP requests. Schemes based on application layer are techniques used for more than just the prediction of the possible throughput – in addition, a number of attempts were proposed on cross-layer throughput estimation. A solution based on machine learning (Support Vector Regression [9]) developed based on [10] is used for the prediction of the possible throughput in [11]. This method trains the throughput prediction via information from the network layer, such as packet loss, delay, and RTT.

Xiong et al. [12] made another argument about the clients' overall perceptions when video quality is changed. They note that the perceived video quality is not easy to describe in an accurate language, or even in other mathematical models that depend on an exact input and output definition. Because of this, they propose using fuzzy logic (called Network-Bandwidth-Aware Streaming Version Switcher), which is divided into three components, namely sensor, controller, and actuator. The sensor works like an estimation resource module, the controller is responsible for the adaptation, and the actuator monitors the controller's decisions. Xiong et al. [12] concluded that the technique was responsive to changes in the network. In [13], fuzzy logic was applied to adapt the video rate to changes in the network's conditions. The proposed algorithm aims to avoid buffer overflow and unnecessary fluctuations in video quality. However, it also suffers from a large variation during changes in video quality. In order to overcome this limitation, Sobhani et al. [14] proposed an AIMD-like fuzzy controller. Their proposed method focuses on both the buffer occupancy and the estimated throughput. Nevertheless, fuzzy logic would require domain expert knowledge, which can be difficult to acquire.

Chien et al. [15] proposed the use of a decision tree in order to map network-related features onto the video rate. Instead of introducing a new algorithm, a scheme was chosen to improve the accuracy of existing adaptation algorithms. In short, these researchers trained the model via a given dataset [16]. The classifier then predicts the current/future requests. Hence, the training process can be applied either on-line or off-line, and the results showed improved prediction accuracy.

Bhat and Bhadu [17] proposed a machine learning approach for video streaming with DASH to help clients adapt to changes in the streaming environment. Their objective was to allow clients to learn about the environment in an unsupervised manner, since they assumed that by doing this, the redundancy of adaptation would be eliminated. They concluded that they could improve QoE and efficiency of bandwidth utilization by up to 68.5%. Van der Hooft et al. [18] used the mean and the average absolute difference in bandwidth, whereas Claeys et al. in [19,20] used the changes information about both the available bandwidth and the buffer state. Then, an agent works by changing the video rate to gradually increase its reward, such as enhancing the Mean Opinion Score (MOS) and decreasing the re-buffering [18]. Martín et al. [21] presented an adaptation algorithm based on the Q-Learning method, called DASH-QL. This approach is based on reinforcement learning [22], which allows clients to learn via experience. This method was able to perform an ideal control like other approaches such as DASH-SDP, while still maintaining adaptivity [21].

Due to long delays and fluctuations in bandwidth, the user's QoE needs to be enhanced constantly. One possible solution involves adaptation logic improvements, which would decrease re-buffering events when transmitting higher video qualities. Bezerra et al. [23] presented a control system (DBuffer) to improve the QoE while a streaming session simultaneously occurs in mobile networks. Their proposed method approximates the most appropriate video quality for specific requests based on the clients' buffer state, which is the main component of their proposed method that controls the adaptation logic [23]. They concluded that their proposed method showed the lowest stall events when comparing the conventional and PANDA adaptation approaches.

## 3    Proposed Method

This section proposes a new approach that utilises machine learning algorithms for determining the best possible quality (in terms of bitrate) that a video streaming client could enjoy when video content is streamed from a DASH server. Current bitrate, buffer state, and the available bandwidth are used in order to determine the optimal target quality of the server, thus ensuring the best possible playback at the client and enhancing the QoE.

**Table 1.** DASH-server and DASH-client hardware specification

|                  | DASH-server                  | DASH-client            |
| ---------------- | ---------------------------- | ---------------------- |
| Processor model  | Intel Xeon E3-1245           | Intel Core i5-4570     |
| Processor speed  | 3.40 GHz (8 cores)           | 3.20 GHZ (4 cores)     |
| RAM              | 7.5 GB                       | 4 GB                   |
| Operating system | Linux Ubuntu-14.04 LTS 64-bit | MS Windows 8.1- 64-bit |

### 3.1    Experimental Setup and Data Acquisition

A DASH-server and a DASH-client were used in order to stream video sequences and acquire network and quality statistics. The hardware specifications of the server and the client system used in this study are summarised in Table 1. An Apache HTTP server was used for the DASH-server system, while MP4Client, an open-source multimedia player from GPAC [24], was used as the DASH-client

**Table 2.** Video sequences used

| Sequence name | Big_Buck_Bunny, Elephants_Dream, Site_sings_the_blues |
|---|---|
| Resolution | 1920 × 1080 pixels |
| Frame rate | 24 fps |
| Duration | 5.45 min (345 s) |
| Segment size (*sec*) | 10, 15 |
| Number of segments | 23 (15 s segments), 35 (10 s segments) |
| Bitrate (kbps) | 125, 250, 375, 500, 750, 1000, 1500, 2000 |

at the client system, since it allows clients to retrieve the most appropriate video content based on the given configurations. Furthermore, the bandwidth was capped at 2.5 Mbps using the NetEm emulator [25] in order to create a realistic test environment for the DASH streaming system.

Raw video encoding is achieved by using the FFmpeg command line tool to produce different versions of each video sequence encoded with the H.265/HEVC codec [26] at the following eight average bitrates: 125, 250 375, 500, 750, 1000, 1500, and 2000 kbps, keeping the original resolution and frame rate. Then, the encoded videos are chunked into smaller segments of a pre-set duration using MP4Box [27]. In this study, segments of 10 and 15 s were created for each of the encoded video sequences. The produced segments are kept on the DASH-server for future retrieval by the DASH-clients. After dividing the encoded video sequences into segments, the Media Presentation Description (MPD) is created automatically for future use by the system. Every available representation of each video sequence (different bitrates, frame rates, resolutions, segments, codecs, etc.) is recorded on the MPD file. The DASH-client then uses the MPD file to choose the most appropriate video representation through HTTP-GET or Partial GET requests, based on the bandwidth conditions available during streaming and the DASH-client's buffer state. An overview of the MPEG-DASH system used in this study is provided in Fig. 1.

Three raw high definition (1080p at 24 fps) video sequences [28] with no copyright restrictions were used in this study. The duration of each video sequence was 5.45 min (345 s) and each sequence was encoded and segmented as previously described. Details about the video sequences used are provided in Table 2. Then, simulating a Video on Demand (VoD) scenario, the examined video sequences were streamed to the DASH-client using the DASH-server, targeting the maximum possible quality.

## 3.2   Feature Extraction and Classification

Information about the streamed content was retrieved by the DASH-client for all the available representations of the three examined video sequences (average bitrate and segment size). For each segment of each representation,

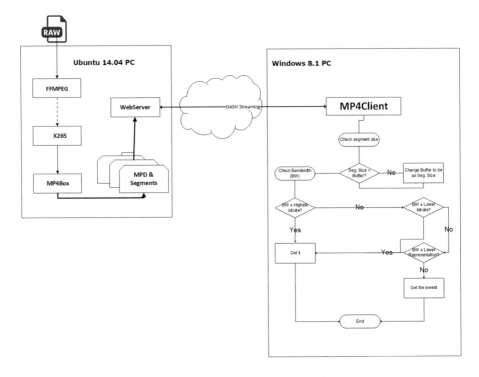

**Fig. 1.** Diagram of the examined MPEG-DASH system

the following information was recorded: the bitrate of played video in *bps* (not the original bitrate but the one served to the client), the actual available bandwidth in *bps*, and the DASH-Client's buffer size in *bytes*. Each triplet (bitrate, bandwidth, buffer size) was then used as a feature vector labelled with the original average bitrate of the examined segment (one of 125, 250 375, 500, 750, 1000, 1500, and 2000). This process led to 280 feature vectors (samples) for each video sequence when segmented with a segment duration of 10 s, and to 184 feature vectors for each video sequence segmented with a segment duration of 15 s. Finally, two set were created, one referring to 10 s segments that had 840 samples and one referring to 15 s segments that had 552 samples, with each sample belonging to one of the eight quality (bitrate) categories. Linear and non-linear classifiers were then used in order to create a machine learning model that can predict the quality class. The motive behind this approach is to assist the DASH-client in targeting the optimal quality level depending on the available resources of the client.

## 4    Experimental Results

Multi-class supervised classification experiments were conducted in order to evaluate the performance of the proposed model. The examined classification algorithms were the $k$-Nearest Neighbour ($k$-NN) for $k = 1, 3, 5, 10$, the Linear Support Vector Machines (SVM), the SVM with RBF and quadratic kernels, the Decision Trees (DT), and Linear Discriminant Analysis (LDA). The one vs one method was used in order to achieve multi-class classification using the SVM classifiers. Furthermore, features were normalised before the classification experiments due to their difference in range, and 10-fold cross validation was employed in order to avoid over-fitting the examined models. The available MATLAB [29] implementations were used for all the classifiers. Results in terms of classification accuracy for each segment size are shown in Table 3. It is evident that the SVM with the quadratic kernel achieves the highest accuracy for both the examined segment sizes, reaching a 79% accuracy for video sequences with a 10 s segment size, and 68.1% for a 15 s segment size.

**Table 3.** Classification results for each segment size

| Classifier | Accuracy (%) | |
|---|---|---|
| | 10 s | 15 s |
| 1-NN | 71.2 | 61.4 |
| 3-NN | 74.0 | 62.0 |
| 5-NN | 75.2 | 62.3 |
| 10-NN | 74.3 | 58.3 |
| SVM (Linear) | 76.5 | 47.6 |
| SVM (RBF) | 77.1 | 63.6 |
| SVM (Quadratic) | **79.0** | **68.1** |
| Decision Trees | 78.6 | 55.8 |
| LDA | 66.1 | 42.0 |

## 5    Conclusion

In this work, we proposed and evaluated a machine learning approach for selecting the optimal streaming quality for an MPEG-DASH video streaming server. Client-side features that included the current bitrate, the current bandwidth, and the current buffer size were used in order to create a machine learning model that is able to distinguish the target quality level in terms of the average bitrate used for encoding the video sequence with the H.265/HEVC codec. Supervised classification experiments showed that the SVM classifier with quadratic kernel function provided the highest classification accuracy, which reached 79% when the video segment size is set to 10 s and 68.1% when set to 15 s. As a result, the proposed model could be successfully utilised within an MPEG-DASH client for selecting the optimal quality for each client based on the client's current capabilities, leading to enhanced overall QoE. Future work could evaluate the proposed approach for different segment sizes (e.g. 1, 2, 5 and 20 s), as well as additional features related to the client's capabilities.

# References

1. Cisco: Cisco visual networking index: Global mobile data traffic forecast update (2017). https://www.cisco.com/c/en/us/solutions/collateral/service-provider/visual-networking-index-vni/mobile-white-paper-c11-520862.html. White Paper. Accessed 23 Jan 2018
2. Microsoft Corporation: [MS-SSTR]: Smooth Streaming Protocol. V8.0-2017/09/15 (2015). https://winprotocoldoc.blob.core.windows.net/productionwindowsarchives/MS-SSTR/[MS-SSTR].pdf. White Paper. Accessed 2 Oct 2017
3. Adobe Systems Inc.: High-quality, network-efficient HTTP streaming. https://www.adobe.com/products/hds-dynamic-streaming.html. Accessed 21 Jun 2017
4. Pantos, R., May, W.: HTTP Live Streaming. RFC 8216, RFC Editor (2017)
5. ISO/IEC: Dynamic adaptive streaming over HTTP (DASH) - Part 1: Media presentation description and segment formats. Standard, International Organization for Standardization (2014)
6. Oyman, O., Singh, S.: Quality of experience for http adaptive streaming services. IEEE Commun. Mag. **50**(4), 20–27 (2012). https://doi.org/10.1109/MCOM.2012.6178830
7. Petrangeli, S., Wu, T., Wauters, T., Huysegems, R., Bostoen, T., Turck, F.D.: A machine learning-based framework for preventing video freezes in HTTP adaptive streaming. J. Netw. Comput. Appl. **94**, 78–92 (2017). https://doi.org/10.1016/j.jnca.2017.07.009
8. Chan, K.M., Lee, J.Y.B.: Improving adaptive HTTP streaming performance with predictive transmission and cross-layer client buffer estimation. Multimedia Tools Appl. **75**(10), 5917–5937 (2016). https://doi.org/10.1007/s11042-015-2556-y
9. Cortes, C., Vapnik, V.: Support-vector networks. Mach. Learn. **20**(3), 273–297 (1995). https://doi.org/10.1007/BF00994018
10. Mirza, M., Sommers, J., Barford, P., Zhu, X.: A machine learning approach to TCP throughput prediction. IEEE/ACM Trans. Netw. **18**(4), 1026–1039 (2010). https://doi.org/10.1109/TNET.2009.2037812
11. Tian, G., Liu, Y.: Towards agile and smooth video adaptation in HTTP adaptive streaming. IEEE/ACM Trans. Netw. **24**(4), 2386–2399 (2016). https://doi.org/10.1109/TNET.2015.2464700
12. Xiong, P., Shen, J., Wang, Q., Jayasinghe, D., Li, J., Pu, C.: NBS: a network-bandwidth-aware streaming version switcher for mobile streaming applications under fuzzy logic control. In: 2012 IEEE First International Conference on Mobile Services, pp. 48–55 (2012). https://doi.org/10.1109/MobServ.2012.10
13. Vergados, D.J., Michalas, A., Sgora, A., Vergados, D.D.: A control-based algorithm for rate adaption in MPEG-DASH. In: IISA 2014, The 5th International Conference on Information, Intelligence, Systems and Applications, pp. 438–442 (2014). https://doi.org/10.1109/IISA.2014.6878834

14. Sobhani, A., Yassine, A., Shirmohammadi, S.: A fuzzy-based rate adaptation controller for DASH. In: Proceedings of the 25th ACM Workshop on Network and Operating Systems Support for Digital Audio and Video, NOSSDAV 2015, pp. 31–36. ACM, New York (2015). https://doi.org/10.1145/2736084.2736090

15. Chien, Y.L., Lin, K.C.J., Chen, M.S.: Machine learning based rate adaptation with elastic feature selection for HTTP-based streaming. In: 2015 IEEE International Conference on Multimedia and Expo (ICME), pp. 1–6 (2015). https://doi.org/10.1109/ICME.2015.7177418

16. Basso, S., Servetti, A., Masala, E., De Martin, J.C.: Measuring DASH streaming performance from the end users perspective using Neubot. In: Proceedings of the 5th ACM Multimedia Systems Conference, MMSys 2014, pp. 1–6. ACM, New York (2014). https://doi.org/10.1145/2557642.2563671

17. Bhat, A.R., Bhadu, S.K.: Machine learning based rate adaptation in DASH to improve quality of experience. In: 2017 IEEE International Conference on Smart Technologies and Management for Computing, Communication, Controls, Energy and Materials (ICSTM), pp. 82–89 (2017). https://doi.org/10.1109/ICSTM.2017.8089131

18. van der Hooft, J., Petrangeli, S., Claeys, M., Famaey, J., Turck, F.D.: A learning-based algorithm for improved bandwidth-awareness of adaptive streaming clients. In: 2015 IFIP/IEEE International Symposium on Integrated Network Management (IM), pp. 131–138 (2015). https://doi.org/10.1109/INM.2015.7140285

19. Claeys, M., Latré, S., Famaey, J., Wu, T., Van Leekwijck, W., De Turck, F.: Design of a Q-learning-based client quality selection algorithm for HTTP adaptive video streaming. In: Adaptive and Learning Agents Workshop, part of AAMAS2013, Proceedings, pp. 30–37 (2013)

20. Claeys, M., Latre, S., Famaey, J., Turck, F.D.: Design and evaluation of a self-learning HTTP adaptive video streaming client. IEEE Commun. Lett. **18**(4), 716–719 (2014). https://doi.org/10.1109/LCOMM.2014.020414.132649

21. Martín, V., Cabrera, J., Garca, N.: Evaluation of Q-learning approach for HTTP adaptive streaming. In: 2016 IEEE International Conference on Consumer Electronics (ICCE), pp. 293–294 (2016). https://doi.org/10.1109/ICCE.2016.7430618

22. Sutton, R.S., Barto, A.G.: Reinforcement Learning: An Introduction. MIT Press, Cambridge (1998)

23. Bezerra, D., Ito, M., Melo, W., Sadok, D., Kelner, J.: DBuffer: A state machine oriented control system for DASH. In: 2016 IEEE Symposium on Computers and Communication (ISCC), pp. 861–867 (2016). https://doi.org/10.1109/ISCC.2016.7543844

24. GPAC: Osmo4 / MP4Client, a powerful multimedia player. https://www.gpac-licensing.com/gpac/. Accessed 3 Feb 2017

25. The Linux Foundation: netem wiki. https://wiki.linuxfoundation.org/networking/netem. Accessed 24 Apr 2017

26. ITU-T: H.265: High efficiency video coding. ITU-T Rec. H.265 (2016)

27. GPAC: MP4Box General Documentation. https://gpac.wp.imt.fr/mp4box/mp4box-documentation/. Accessed 3 Feb 2017

28. Xiph Foundation: Xiph.org video test media [derf's collection]. https://media.xiph.org/video/derf/. Accessed 21 Jan 2017

29. Mathworks: Matlab 2016b (2016). https://uk.mathworks.com/products/matlab.html

# Analysis of Traffic Accident in Riyadh Using Clustering Algorithms

Alaa Almjewail, Aljoharah Almjewail, Suha Alsenaydi, Haifa ALSudairy, and Isra Al-Turaiki[✉]

College of Computer and Information Science, King Saud University, Riyadh, Saudi Arabia
{436204482,437203996}@student.ksu.edu.sa,
{aalmjawel,salsenaidy}@su.edu.sa, ialturaiki@ksu.edu.sa

**Abstract.** Saudi Arabia is considered one of the top ranking countries in terms of car accident rates. Riyadh, the capital of Saudi Arabia, has the highest rate of accidents, reaching 29.20%. This study aims to analyze traffic accident records in Riyadh and attempt to determine the locations with high risk of accidents. In this paper, we apply data mining techniques in order to understand traffic accidents characteristics and help identify black spots. The two clustering techniques that are used are: k-means and DBscan. Results revealed the most occurring locations for accidents.

**Keywords:** Data mining · Road accident · Riyadh · K-means
DBscan · Black spot · Hot spot

## 1 Introduction

Road traffic accidents continue to be a major problem worldwide. According to the *World Health Organization*, every year 1.25 million people die in road accidents [1]. In Saudi Arabia, road accidents rates continue to grow. According to the *Secretary General of Traffic Safety Committee* in the eastern province, Eng. Sultan Al-Zahrani, "Saudi Arabia witnesses up to 7,000 deaths and over 39,000 injuries, annually". At least one traffic accident occurs in Saudi Arabia every minute. Saudi Arabia ranks at the top of the list of the world's countries with highest total traffic deaths per 100,000 persons (about 21 deaths) [2]. This issue has caught the attention of both government authorities and the public. Traffic accidents is a huge problem not only causing the loss of lives, but also putting a burden on the economy.

In this paper, we focus on analyzing traffic accidents data obtained from the *General Department of Traffic* of Saudi Arabia. This work is a first step towards the identification of road accident black spots in Riyadh. *Black spots* refer to high risk accident locations [33]. In another definition, black spots are any locations that have a higher number of crashes than other similar locations

© Springer International Publishing AG, part of Springer Nature 2018
M. Alenezi and B. Qureshi (Eds.): *5th International Symposium on Data Mining Applications*, pp. 12–25, 2018.
https://doi.org/10.1007/978-3-319-78753-4_2

as a result of local risk factors [32]. The identification of black spots provides a better understanding of the traffic accident causes and help decision makers put effective measures in place. We hope that the results we obtain can contribute to the reduction of the number of accidents and make the roads safer to commuters. We apply data mining techniques, in particular, the two clustering algorithms: *k-means* [31] and *DBSCAN* [30] to understand accident characteristics and for finding black spots.

The rest of the paper is organized as follows: Sect. 2 discusses some of the related work in the identification of black spots using data mining techniques. Section 3 describes our methodology, including data preprocessing and the data mining algorithms used. The experiment's results and discussion are presented in Sect. 4. Finally, Sect. 5 presents the conclusion of this paper.

## 2   Literature Review

An intensive accident distribution section of the road is called accident-prone point or accident-prone section. In these sections, the probability of an accident is much higher than the average road per kilometer of the time unit [15]. The purpose is to determine if specific location (point, section, region) are at the risk of danger or not, which may have the possibility of an accident happening and causing heavy loss. Determining of road black spots usually considers the numbers of the traffic accident (frequency) and lose (severity) [15].

Traffic safety management is one of the main responsibilities of governments. It is important to investigate and understand the factors and attributes leading to car accidents. Data mining can be used in order to identify patterns and predict future behavior. This leads to better decisions for reduced numbers of road accidents [3]. Here, we discuss some of the literature related to determining traffic accident black spots.

Kaur and Kaur [4] predicted locations of black spots on State Highways and Ordinary District Roads. The study determined the accidental attributes, and factors to reduce the risk of accidents. For analyzing incidental data, they used exploratory visualization techniques and machine learning algorithms. The machine learning includes KNN Algorithm and K-means Algorithm. The aim and purpose of the study is to helps derive the statistical model by using data mining algorithms and exploratory visualization techniques. Since the study, it has been concluded from all the techniques applied that most of the accidents on State Highways occur mostly on Straight Roads and Ordinary District Roads majorly occur at Cross intersection.

Thapa and Lee [5], used Association Rule Mining and Clustering algorithm in order to help realize the traffic violation patterns and black spots of traffic violations. They looked into K-means clustering with some improvements to aid the process of identification of patterns and black spots. The analysis showed that by generating association rules the identification of accident circumstances that frequently occur together is facilitated. This leads to a strong contribution towards a better understanding of the occurrence of traffic accidents. However, association rules do describe the co-occurrence of accident circumstances, but they do not give any explanation about the causality of these accident patterns. The results indicate that the use of the association algorithm is not only allowed to give a descriptive analysis of accident patterns on high-frequency accident locations, it also creates the possibility of finding characteristics of the accident that distinguish between high-frequency accident locations and low-frequency accident locations.

Kumar and Toshniwal [6] used data mining techniques to analyze traffic accidents data in India. They determined the road accidents that are more frequent at particular locations and extracted characteristics of accidents by using association rule mining that helped them identify the connections between attributes of accidents. Thus, identified specific features of road accidents that lead to incidents happening frequently in these locations. Similarly, they used k-means clustering algorithm to classify the accident locations into high frequency, moderate-frequency and low-frequency accident locations. As a result of the study, the researchers mentioned that their approach was quite sufficient to uncover reasonable information from the selected dataset, but the attributes were not enough to extract the best features of accident occurrence.

In Mashhad, a city in Iran, the researchers attempted to examine and compare several kinds of traffic accidents in terms of spatial attributes. They combined the geo-information technology and spatial-statistical analysis. The goal was to examine 4 clustering analyses to better understand traffic accident patterns in the complex urban network. Geographic Information System technology is a common tool for analysis of black spots and visualization of accident data. The fundamental objective of this study was to inspect the distribution of accidents through identification of black spots using GIS and spatial statistics. The analysis process used many methods and techniques, such as , traditional kernel density estimation, nearest neighbor analysis, and K-function. Nearest neighbor analysis and K-function were to examine the presence or absence of clusters of accidents. In this paper, they used ArcGIS1 and SANET2 software, ArcGIS to analyze kernel density and SANET for spatial analysis on the network and nearest neighbor [7].

Yalcin and Duzgun [8] employed Geographic Information Systems. Three methods of spatial accident pattern analysis on a network (kernel density, nearest neighbor distance, and the K- function) were used. It focused on the numbers of the accidents according to vehicle types, especially on two-wheeled vehicles. The analyses showed that accidents are clustered rather than happening by chance. The clusters refer to the locations where incidents occur intensely. The results show two-wheeled vehicle accidents have a big ratio of accidents. Furthermore, they found that the accidents involving two-wheeled vehicles are clustered on the main transportation road.

In their study, Dereli and Erdogan [9] considered determining the traffic accident black spots through the use of GIS-aided spatial statistical methods. According to the authors, an ideal descriptive model for determining the traffic accident black spots should use three spatial statistical methods: Poisson regression, negative binomial regression, and empirical Bayesian method. Additionally, they suggest that the identification of black spots should depend on the Geographic Information Systems (GIS), which is greatly used in the modern era. The study examined the viability of their model by analyzing the traffic accidents. The study concludes that empirical Bayesian method is the most suitable method for accident black spot detection studies since it yields more accurate and precise results.

Past studies have focused on developing suitable measures that can be used to measure the safety status of specific sites. In their research, Hussien and Eissa [10] sought to identify hazardous road locations in Saudi Arabia. They used descriptive statistical analysis to determine the general trend of traffic accidents and ascertain dangerous roadways in KSA network. They utilized formal statistical methods and procedures such as Poisson regression analyses and were able to reach statistical inference on the population from the sample. To point out the hazardous locations, the structure of the underlying road networks was considered as well as the number of accidents, injuries, and deaths. They attributed the upward trajectory to the increase in the number of registered vehicles and the construction and rehabilitation of paved roads that encouraged increased speeds. When the number of traffic accidents inside cities is compared to that outside the cities, they found accidents that occurred outside the cities are significantly more, excluding Riyadh, Jeddah, and Jazan. They attributed this phenomenon to the increase of vehicle speeds on the highway and overcapacity [10].

Martin et al. [11] conducted a study focusing on the use of data mining techniques to road safety improvement on Spanish roads. Martin et al. assert that data mining techniques are employed to find hidden relationships between the characteristic of the road, ESM(Susceptible Elements for Improving), and crashes. They recommend the use of conceptual approach which enables extraction and a given algorithm to implement the appropriate data mining technique. The study aimed to correlate ESM, crashes and TCA, a consideration is given to decision tree to obtain relations that are understandable.

In Morocco, a new approach had been proposed to reduce the number of rules that helped the police decision makers to choose the suitable rules for their need. The researchers used integration technique to achieve their goal. This technique is divided into two steps the first step is using association rules, they especially chose Apriori algorithm to extract the association rules. Apriori algorithm is applying minimum support to find all frequent item sets. The second step is using multi-criteria analysis approach to assign the rules to the relevant category. However, the researchers faced a problem when they had to assign the result to the most relevant category then they resorted to Electra Tri method to solve this problem. The researchers observed that when they applied the Multi-criteria analysis especially the Electre Tri method, they obtained the 12 best rules after ignoring the redundant and non-interesting rules where as the Apriori algorithm extracted fourteen rules [12].

In Montella [13] research, seven commonly applied black spots identification (HSID) methods were compared against four robust and informative quantitative evaluation criteria. The following HSID methods were compared:

crash frequency (CF), equivalent property damage only (EPDO) crash frequency, crash rate (CR), proportion method (P), empirical Bayes estimate of total-crash frequency (EB), empirical Bayes estimate of severe-crash frequency (EBs), and potential for improvement (PFI). The HSID methods were compared using the site consistency test, the method consistency test, the total rank differences test, and the total score test. Quantitative evaluation tests showed that the EB method performed better than other methods. The test results highlights that the EB method is the most consistent and reliable method to determine priority investigation locations. Yet, the results of the study, together with the results of previous research, strongly suggest that the EB method should be standard in determining black spots.

Szénási et al. [14], Clustering Algorithm is a method of identifying black spots using algorithms. DBSCAN is a high-performance algorithm that uses arbitrary shape to classify data and handles noise efficiently. It is primarily based on two algorithms, one that brings data into the DBSCAN and the other operates data outside the DBSCAN. The performance of DBSCAN has been found to be extremely good, dependent on the choice of input. This is because it mainly relies on two contributions that are the maximum radius and a minimum number of points required to form clusters. Black spots are identified by seemingly sophisticated methods of clustering points.

Ye et al. [15] produced system based on GIS technology to deal with urban road traffic accident information. The system had many modules such as traffic accident information module. This module analyzed accident black spots and it used two methods clustering analysis DBSCAN algorithm and thematic map analysis. DBSCAN (density-based clustering algorithm) was used to cluster points of two-dimensional planes. When the system clustering analyzes the black spots, they should set the Eps the radius neighborhood and MinPts the least frequency of accident point. To make judgment of the accident black spot, the clustering analyzed the traffic accident point then calculated the number of density-reachable point at the neighborhood of each point in cycle.

Chen et al. [16] discussed the utilization of geocoding technology to make traffic accidents with spatial coordinate. They used geocoding due to the location traffic accidents were described as address with text this step made the capability to display the data on the map. Also, they presented the method that took potential of reducing accident as index to extract black spots. Based on the data used in this research, the intersection accidents and microscopic road section accidents were grouped into several categories by the type of traffic facilities. After that, they calculated the difference between the number of the accident and the average number of the accidents which happened in the same kind of traffic facilitates, this value is called the potential of reducing accidents. Then, arranged the values in descending order and the traffic facilities were considered as black spots.

Hosseinlou et al. [17] tried to find effective results in identifying traffic black spots by using adaptive neuro-fuzzy inference system. This system adopted the ambiguous and complicated data and adjusted the required parameter and components for predicting other data that are not faced with yet. The dataset contained the accidents that occurred in Karaj-Chalus, two-lane major road in Iran during three months. In this research, fuzzy inference system used Sugeno type. And it trained applying hybrid optimization routine back propagating algorithm in combination with a least square type of method and accidents data. Finally, they tested the system by a complete dataset and it can predict 96.85% of accident frequencies, which indicated a good approximation of prediction. In this paper, we choose to utilize two data mining techniques. K-mean and DBscan in order to determine black spot locations of the road accidents.

# 3 Method

## 3.1 Data Understanding and Exploration

We obtained the dataset from the *General Department of Traffic* of Saudi Arabia. We were provided with around 242,000 records of traffic accidents that occurred in Riyadh for three years, 2013, 2014, and 2015. The accident data were in computer-ready form; each year was separated in a spreadsheet format. There were three main table sheets *Accident, Vehicle* and *Parties*. For each accident table, approximately twenty-seven attributes describing accident situation. These attributes include neighborhoods, zones, streets, XY coordinates, time of accident, weather conditions, road surface conditions, lighting conditions, injury severity level, and accident reasons. The Vehicle table sheet, contained all information that is related to vehicles, such as, vehicle manufacturer, vehicle model, color and percentage of error. The Parties table describes people involved in the accident, it includes license type, gender, health status and nationality.

## 3.2 Data Preparation and Pre-processing

Data Preparation is the process of collecting, cleaning, and consolidating data into one file or data table, primarily for use in analysis. This step usually takes 80% of the time in the data analytics life cycle. This step includes: filling in missing values, normalizing data, and/or integrating multiple data sources. As mentioned above, we obtained three years of traffic accidents dataset, each one is represented by three tables. We chose to use the Accidents tables, which is more relevant to our research aim.

Some attributes were discarded. For example, the attribute *Accidents Report Time* had missing values for a large number of records. Other attributes, such as *Weather* had the same value for 97% of records, as shown in Fig. 1.

**Fig. 1.** Values for attribute *Weather*

We noticed that the accident locations, represented as XY coordinates, could not be represented on *Google Earth*. Thus, pre-processing for this attribute was required. We searched for the type of system used for the XY coordinates in the original dataset. We were able to identify that the coordinate system used was UTM of type "WGS 1984 UTM Zone 38N". We used ArcMap [20, 21] software to convert the coordinates of the type UTM to Decimal Degrees. Finally, ten attributes were kept for 246,814 records of accidents. Kutool [18] was used in order to integrate the three Microsoft Excel sheets into one. We separated the column that contains time and date into two columns, one for time and another for date. After that, we utilized Orange tool [19] in order to fill missing values with average values, converting numerical values into categories, as well as excluding instance with meaningless values corresponding to any feature as shown in Fig. 2. The ten attributes and their possible values are described in Table 1.

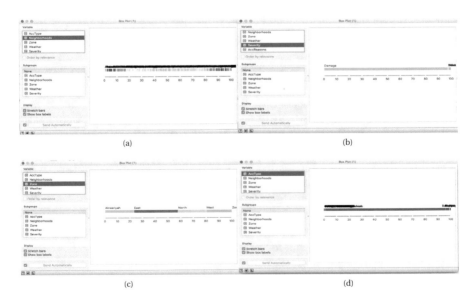

**Fig. 2.** Dataset feature.

## 3.3  Traffic Accident Black Spot Analysys Techniques

Clustering is the process of partitioning a set of records into groups. Each individual group, called *cluster*, contains records that are similar to one another and different from records in other clusters. Clustering is a key data mining task, and it is usually referred to as *unsupervised learning*. Unlike classification and regression, clustering is used to analyze unlabeled dataset with prior knowledge about these labels. Clustering is used in many applications in various fields, including: customer segmentation, image recognition, and Web search [34]. In this paper, we apply two well-known clustering algorithms: k-means and DBSCAN. In this section, we give a brief description of these two algorithms.

**DBSCAN Algorithm.** The *Density-Based Spatial Clustering of Applications with Noise*, is a density-based clustering algorithm. The basic idea is to find records that have dense neighborhoods. The neighborhood size, $\epsilon$, is fixed and

**Table 1.** Attributes description

| Feature description | Value |
|---|---|
| Type of accident | Bumped a moving vehicle |
| | Bumped a traffic signal |
| | Bumped a vehicle parked |
| | Bumped animal |
| | Bumped bicycle |
| | Bumped body fixed |
| | Bumped electricity tower |
| | Bumped the barrier |
| | Bumped through the fence |
| | Bumped waste container |
| | Bumped with the panel |
| | Bumped with a tree |
| | Burning vehicle |
| | Coup vehicle |
| | Fall from a cliff |
| | Fall from the bridge |
| | Motorcycle bumped |
| | Run over |
| Accident zone | North |
| | East |
| | West |
| | Middle |

(continued)

**Table 1.** (*continued*)

| Feature description | Value |
|---|---|
| Accident reasons | Attached to a vehicle |
| | Contrary to the board's traffic |
| | Cross the red signal |
| | Defect in, steering the vehicle |
| | Defect with, electricity |
| | Defective Tires |
| | Do not leave enough distance |
| | Downhill |
| | Drifting |
| | Fatigue |
| | Lack of lights |
| | Overloaded |
| | Parking violation |
| | Play in the street |
| | Preoccupation on driving |
| | Reversing traffic |
| | Rise before stopping |
| | Sitting in a box vehicle |
| | Sleep |
| | Slippery.peeding |
| | Sudden deviation |
| | The descent before stopping |
| | The lack of warning signs |
| | Under the influence of heady |
| | Unprotected work area |
| | Violation priority |
| | Walk in the middle of the street |
| Street | Number of streets that have accidents |
| Degree, of accident severity | No injury |
| | Injury |
| | Death |
| XCoordinate | The horizontal value in a pair of coordinates: how far along the point is |
| YCoordinate | The vertical value in a pair of coordinates: how far up or down the point is |
| Neighborhoods | The small area in zone 200 neighborhoods |
| Time | 1–24 |
| Dates | dd-mm-2013 -2015 |

defined by the user. DBSCAN relies on another user-defined parameter, *MinPts*, which indicates the density threshold of clusters. An object in the dataset is called a *core object* if there are at least *MinPts* neighbors in its $\epsilon$ neighborhood. Core objects are the indicators of dense regions. DBSCAN has attracted a lot of research interest during the last decades with many extensions and applications. Compared to other clustering algorithms, DBSCAN has many attractive benefits, including: ability to detect clusters with arbitrary shape and to deal with outliers [23,24]. In addition, the significant advantage of DBSCAN algorithm is that its clustering speed is fast. Likewise, it effectively deals with noise points and discovers spatial clustering of random shape [15].

The DBSCAN algorithm can be summarized in the following steps [25]:

- Find the radius $\epsilon$ neighbors of every point, and identify the *core points* with more than given threshold *MinPts* neighbors.
- Find the connected components of core points on the neighbor graph, ignoring all non-core points.
- Assign each non-core point to a nearby cluster at most distance radius $\epsilon$, otherwise assign it to noise (in our case noise is an accidents that does not belong to a clusters).

**K-means Algorithm.** k-means is a partitioning-based clustering algorithm. It is based on grouping objects into clusters and calculating the value of a *centroid*. The cluster centroid is the mean value of the records in the cluster. The final clusters produced by k-means need to have high intra-cluster similarity and less inter-cluster similarity. The number of clusters, $k$, is given by the user. k-means algorithm is simple and commonly used clustering algorithms in industrial and scientific areas. It can deal with numerical data and is robust for cluster analysis [22]. K-means clustering algorithm includes the following steps [26,27]:

- Randomly, generates $k$ points, representing the initial cluster centroids.
- Each object assigns to the closest centroid based on the squared Euclidean distance.
- When all data points have been assigned, the centroids are recalculated.
- Steps 2 and 3 are repeated until no objects change clusters and the centroids no longer moves, the sum of the distances is minimized, or some maximum number of iterations is reached.

## 4   Experimental Results

In this section, we show the result of using Weka [28] and Rapidminer [29] tools to cluster the dataset. We then visualize the records on Google map by using ArcMap tool [20].

## 4.1   K-means Algorithm

In order to understand the characteristics of accidents, we apply k-means using Weka to the dataset, excluding XY coordinates. We tried three values for k, 5,15,17. The results show that "Azezeah" neighborhoods was a common place where traffic accidents occur. Mostly, traffic accidents happened in the evening and the majority occurred in 2015. Accidents reasons ranged between speeding and preoccupation on driving. Overall, we observed that accidents are concentrated in the north and east of Riyadh. We believe the reason behind this is that most of the institutions, companies, and government agencies are located there. Table 2 shows the running time in seconds and the number of iterations for each value of $k$.

**Table 2.** K-means clustering setting

| K-means values | Distance measure | Time taken to build model | Number of iterations |
|---|---|---|---|
| 5 | Euclidean | 0.95 s | 4 |
| 15 | | 2.21 s | |
| 17 | | 2.41 s | |

## 4.2   DBSCAN Clustering

We used RapidMiner to conduct the entire process. The main difficulty in applying DBSCAN is setting the two parameters $\epsilon$ and *MinPts*. We experiment with various settings as shown in Table 3. We focus on the second situation in the table. DBSCAN generated 36 clusters. Most of the accidents occur in East and North of Riyadh, where most of them belong to cluster 36. Figure 3 shows the number of accidents in each cluster. Figure 4 shows the clustering in two dimensional space X and Y.

**Table 3.** DBSCAN parameter settings and run time in hours

| Min points | Epsilon | Distance measure | Time of process |
|---|---|---|---|
| 3 | 500 | Euclidean | 7:25:34 |
| 0.0045 | 50 | | 6:32:15 |
| 1.0 | 5 | | 6:17:23 |

## 4.3   Visualization

We used ArcMap software to represent the coordinate on Google Earth as shown in Fig. 5. The coordinates of the accident locations were represented over three years in Riyadh (2013, 2014 and 2015).

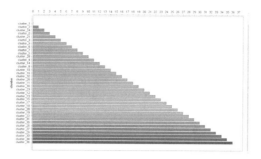

**Fig. 3.** Accident frequency in each cluster

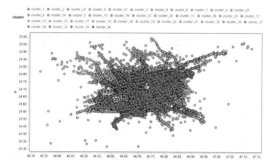

**Fig. 4.** The clustering of the dataset

**Fig. 5.** Visualization the coordinates in Google map.

## 5   Conclusion

Traffic accidents continue to be a major problem Riyadh, Saudi Arabia. In this study, we attempt to understand traffic accident data using clustering techniques. Two commonly used data mining techniques were applied: k-means and DBScan. k-means was applied to understand accident characteristics. DBSCAN was applied to identify black spots. The results indicated that "Azezeah" was the most frequent location where accidents occur. The accidents were centered in the north and east of Riyadh and most of them occurred in the evening. The common reasons behind the accidents were speeding and preoccupation while driving.

This work is still in progress. We plan to continue investigating the applicability of k-means and DBSCAN to cluster geolocation datasets. In addition, we would like to evaluate the quality of the obtained clusters. We intend to identify black spots at street level.

## References

1. WHO — Number of road traffic deaths, WHO. http://www.who.int/gho/road_safety/mortality/traffic_deaths_number/en/. Accessed 12 Dec 2017
2. Traffic accidents occur every minute in KSA, Arab News, 13-Mar-2015. http://www.arabnews.com/saudi-arabia/news/717596. Accessed 11 Dec 2017
3. Ebrahimkhani, S., Begham, B.S., Moradkhani, F.: Road accident data analysis: a data mining approach. Indian J. Sci. Res. **3**(3), 437–443 (2014)
4. Kaur, G., Kaur, H.: Black spot and accidental attributes identification on state highways and ordinary district roads using data mining techniques. Int. J. **8**, 5 (2017)
5. Thapa, S., Lee, J.: Data Mining Techniques on Traffic Violations. University of Bridgeport, CT (2016)
6. Kumar, S., Toshniwal, D.: A data mining approach to characterize road accident locations. J. Mod. Transp. **24**(1), 62–72 (2016)
7. Shafabakhsh, G.A., Famili, A., Bahadori, M.S.: GIS-based spatial analysis of urban traffic accidents: case study in Mashhad. Iran. J. Traffic Transp. Eng. **4**(3), 290–299 (2017)
8. Yalcin, G., Duzgun, H.S.: Spatial analysis of two-wheeled vehicles traffic crashes: Osmaniye in Turkey. KSCE J. Civ. Eng. **19**(7), 2225–2232 (2015)
9. Dereli, M.A., Erdogan, S.: A new model for determining the traffic accident black spots. Transp. Res. Part A **103**, 106–117 (2017)
10. Hussien, H.H., Eissa, F.H.: Identifying hazardous road locations in Saudi Arabia. J. Eng. Technol. Innovation **5**(4), 045–056 (2016)
11. Martin, L., Baena, L., Garach, L., Lopez, G., de Ona, J.: Using data mining techniques to road safety improvement in Spanish roads. Procedia Soc. Behav. Sci. **160**, 607–614 (2014)
12. Addi, A.M., Tarik, A., Fatima, G.: An approach based on association rules mining to improve road safety in Morocco. In: 2016 International Conference on Information Technology for Organizations Development (IT4OD), pp. 1–6 (2016)
13. Montella, A.: A comparative analysis of hotspot identification methods. Accid. Anal. Prev. **42**(2), 571–581 (2010)
14. Szenasi, S., Csiba, P.: Clustering Algorithm in Order to Find Accident Black Spots Identified By GPS Coordinates. In: SGEM 2014. Bulgaria, pp. 497–503, 17-26 June 2014

15. Ye, J., Zhou, Y., Li, M., Wang, C.: Research and implement of traffic accident analysis system based on accident black spot. In: 2010 5th International Conference on Computer Science Education, pp. 1805–1809 (2010)
16. Chen, Y., Wu, H., Liu, C., Sun, W.: Identification of black spot on traffic accidents and its spatial association analysis based on geographic information system. In: 2011 Seventh International Conference on Natural Computation, vol. 1, pp. 143–150 (2011)
17. Hosseinlou, M.H., Sohrabi, M.: Predicting and identifying traffic hot spots applying neuro-fuzzy systems in intercity roads. Int. J. Env. Sci. Technol. **6**, 309–314 (2009)
18. Kutools - Combines More Than 200 Advanced Functions and Tools for Microsoft Excel. https://www.extendoffice.com/product/kutools-for-excel.html. Accessed 15 Dec 2017
19. Orange -Data Mining Fruitful and Fun. https://orange.biolab.si/. Accessed 15 Dec 2017
20. ArcMap — ArcGIS Desktop. http://desktop.arcgis.com/en/arcmap/. Accessed 15 Dec 2017
21. AutoCAD Civil 3D — Civil Engineering Software — Autodesk. https://www.autodesk.com/products/autocad-civil-3d/overview. Accessed 15 Dec 2017
22. Russel, S.J., Norvig, P.: Artificial Intelligence a Modern Approach (2010)
23. Kim, K., Yamashita, E.Y.: Using a k-means clustering algorithm to examine patterns of pedestrian involved crashes in Honolulu, Hawaii. J. Adv. Transp. **41**(1), 69–89 (2007)
24. Kanagala, H.K., Krishnaiah, V.V.J.R.: A comparative study of K-Means, DBSCAN and OPTICS. In: International Conference on Computer Communication and Informatics (ICCCI), pp. 1–6 (2016)
25. Schubert, E., Sander, J., Ester, M., Kriegel, H.P., Xu, X.: DBSCAN Revisited. ACM Trans. Database Syst. **42**(3), 1–21 (2017)
26. Macqueen, J.: Some methods for classification and analysis of multivariate observations. In: Proceedings of Fifth Berkeley Symposium Mathematical Statistics and Probability, vol. 1(233), pp. 281–297 (1967)
27. Trevino, A.: Introduction to K-means clustering 2016. https://www.datascience.com/blog/k-means-clustering. Accessed 25 Nov 2017
28. Weka 3 - Data Mining with Open Source Machine Learning Software in Java. https://www.cs.waikato.ac.nz/ml/weka/downloading.html. Accessed 15 Dec 2017
29. Data Science Platform RapidMiner. https://rapidminer.com/. Accessed 15 Dec 2017
30. Ester, M., Kriegel, H.-P., Sander, J., Xu, X.: A density-based algorithm for discovering clusters a density-based algorithm for discovering clusters in large spatial databases with noise. In: Proceedings of the Second International Conference on Knowledge Discovery and Data Mining, Portland, Oregon, pp. 226–231 (1996)
31. Mac Queen, J.E.: Some methods for classification and analysis of multivariate observations. In: Proceedings of the Fifth Berkley Symposium on Mathematical Statistics and Probability, vol. 1, pp. 281–297 (1967)
32. Elvik, R.: State-of-the-Art Approaches to Road Accident Black Spot Management and Safety Analysis of Road Networks. Report 883. Institute of Transport Economics, Oslo (2007)
33. Geurts, K., Wets, G., Brijs, T.: Profiling High Frequency Accident Locations Using Associations Rules. Steunpunt Verkeersveiligheid (2002)
34. Han, J., Kamber, M.: Data Mining: Concepts and Techniques. The Morgan Kaufmann Series in Data Management Systems, 3rd edn. Morgan Kaufmann, Boston (2011). ISBN 978-0-12-381479-1

# Support of Existing Chatbot Development Framework for Arabic Language: A Brief Survey

Eman Saad AL-Hagbani[(✉)] and Mohammad Badruddin Khan

Information Systems Department, College of Computer and Information Sciences,
Al Imam Mohammad ibn Saud Islamic University (IMSIU), Riyadh, Saudi Arabia
emansaad1989@hotmail.com, mbkhan@imamu.edu.sa

**Abstract.** Technological development is known to support many fields in opening new channels to interact with users and website visitors. Highly intelligent applications such as Chatbots are the future, which are used to add a value to the user's experience in various domains not limited to e-commerce, gaming or education sectors. The main role of these Chatbots is to simulate human conversation to answer the users' questions and fulfill their needs using artificial intelligent algorithms, natural language processing techniques and knowledge base. In recent years, number of different Chatbots were deployed to accomplish different goals in order to help companies to increase their sales and profits. Due to rapid changing environment and increasing demand, Chatbot development frameworks and platforms were created to assist people without any technical background to build a Chatbot easily and rapidly. But in the Arabic world, comparatively less attention has been given to this area where Chatbot applications are still in the beginning stage and immature due to number of reasons including the complexity of Arabic language. This paper presents the review of some existing Chatbot development frameworks and platforms and their support for Arabic language and discusses their support level for the developer in building the Arabic Chatbot as compared to support for other languages like English and French.

**Keywords:** Chatbot · Arabic language · Frameworks · Platforms

## 1 Introduction

Many countries are moving faster toward developing new applications and using new technologies to improve services and deliver them in an effective and efficient manner. In the world of computer sciences, the Artificial Intelligent (AI) tools can help in creating new software solutions and intelligent applications such as human brain simulation, natural language processing, intelligent agents (machine and software) to serve number of areas and sectors.

One of these solutions is a high intelligent application known by Chatbot or Chatterbot. It was created to enhance the human-computer interaction by mimicking a real human conversation through a wide verity of systems and computer programs.

© Springer International Publishing AG, part of Springer Nature 2018
M. Alenezi and B. Qureshi (Eds.): *5th International Symposium on Data Mining Applications*, pp. 26–35, 2018.
https://doi.org/10.1007/978-3-319-78753-4_3

In Dohrmann et al. [1] work, they described the Chatbot as "a server application replying to questions posed via internet in a way that motivates the inquirer to continue the conversation". Basically, this application is created to perform a human-like conversation either text, visual or voice with users via natural language to provide information, answering question or solving problems with the help of some AI algorithms, queries and knowledge base. Some Chatbots are simple as a Q/A (question and answers) systems, while others are powered with AI (Artificial intelligence) which enable them to learn, increase efficacy and handling complex and new queries.

Kurdi et al. [2] pointed out in their study that the architecture of Chatbot can be seen in a Q/A (question and answers) systems, where they divided the system into three main components which are the question classification component, the information-retrieval component and the answer-extraction component. The study also explained that these components are key components in any Q/A systems or dialogue systems such as Chatbots. Since these systems depends on language to interact with users, the difference between the Chatbots is the supported languages(s), where each language includes some distinct features that force these components to work differently in searching, analyzing and retrieving the answers according to the chosen language.

Although the Arabic language is known to be a huge part of the internet content and many mobile and social media application, the Arabic world is lacking the power of the Chatbot application and has less academic attention in contrast with other languages such as English for example. This paper presents a brief review on some of the Chatbot frameworks and platforms status when developing the Arabic Chatbot and supporting the Arabic language.

## 2 Related Work

### 2.1 Chatbot Overview

The Chatbot phenomena started in the early of 1960's when Weizenbaum created the first Chatbot called "ELIZA" designed to simulate psychotherapy to help patient with mental or emotional problems, then in1970 a well know classic Chatbot "PRRY" was created by Colby to simulate a person-centered therapy to understand the paranoid patient [3]. Figure 1 shows a conversation sample with ElZA Chatbot [4].

```
Human:  Men are all alike.
ELIZA:  In what way?
Human:  They're always bugging us about something.
ELIZA:  Can you think of a special example?
Human:  Well, my boyfriend made me come here.
ELIZA:  Your boyfriend made you come here?
Human:  He says I'm depressed much of the time.
ELIZA:  I'm sorry to hear that you are depressed much of the time.
Human:  It's true.  I'm unhappy.
ELIZA:  Do you think coming here will help you not to be unhappy?
Human:  I need some help, that much seems certain.
```

**Fig. 1.** A sample conversation with ELIZA

The development was continued by Fred Roberts when built an award winning Chatbot called Elbot in 2008 that focuses on artificial intelligent by using word synonyms and using the meaning of them in communication [5]. Furthermore, more Chatbots are available to serve different needs now a days. In Chakrabarti and Luger [6] study, they represented the Chatbot's architecture in three main components. As seen in Fig. 2, the first component that connects with the user and capture the user's input is the Chat Interface. Then the Knowledge Engine, which the Chatbot uses to hold specific information about a topic using tags and categories in a basic AIML files, or Topic Hash Table in SQL database. Finally, the main component that manage the process and data flow form input to output which is called the Conversation Engine.

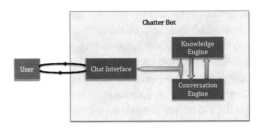

**Fig. 2.** Chatbot architecture

**Chatbot Technologies.** The success of a dialogue application such as Chatbot is based on the type of tools and technologies used in processing and handling natural language. The incorporation of tool handling Natural Language Processing (NLP) which is one of the well-known text mining techniques, aims to ensure the grammatical rules such as the subject-verb and the arrange sentences [7].

In addition, any developer should consider number of known conversational system techniques including [8]:

- SQL queries and Relational Database, to help the Chatbot store large and complex data with the ability of remembering the pervious conversation.
- Parsing and analyzing the user input, using different NLP functions, to fully understand the intention and the meaning of the user's question or needs.
- Ontologies to help the Chatbot compute the relation between the language synonyms, hyponym and other natural language concept names.

Chatbot having the ability to learn from experience and having power to analyze and understand different forms of user's input to grow is a goal for any Chatbot developer and the field of Artificial Intelligent (AI) technology can help a chatbot developer in achieving this goal. Artificial Intelligent (AI) is an area in computer science field and supports the creation of intelligent applications, systems and agents that can think and function just like a human bean. To take advantages form past conversation and chatting experience with users, the AI sub-concept called Machine Learning is used.

Basically, Machine Learning is "the ability of a machine to perform better at a given task using its previous experience" [9]. Therefore, it focuses on the development of Chatbots or conversational agents that can teach themselves new keywords and responses and detect patterns in their knowledge base and adjust their actions accordingly.

Developing a powerful and intelligent application such as Chatbot could be done through two approaches:

1. Creation of a Chatbot from scratch and usage some of the available libraries and programming background to support development process.
2. Creation of a Chatbot using the Chatbot platforms and frameworks that allow to build and develop a Chatbot applications effortlessly and rapidly to assist people without having a programming background or knowledge.

Choosing between them is based on number of factors such as the Chatbot's main goal, the stakeholders, the developing cost or the working field.

## 2.2   The Arabic Language

The Arabic language is known to be "a Semitic language which differs from Indo-European languages syntactically, morphologically and semantically" [10]. Also, the Arabic writing has three long vowels and twenty-five consonants that are written form the left to the right (both can take different shapes depending on the position in the word) and the short vowels that are not a part of the Arabic alphabet but used above or under a consonant which helps to give a different sound and meaning of a word as shown in Fig. 3 examples using the word "علم" [10].

| Arabic | Transliteration | PoS | meaning |
|--------|-----------------|-----|---------|
| عَلَم | 'alam | n | flag |
| عِلْم | 'ilm | n | science |
| عُلِمَ | ulima | v | known |
| عَلَّمَ | 'allama | v | teach |
| عَلَم | 'alam | a | famous |

**Fig. 3.** The Arabic short vowels

In the Arabic language, there are two main classifications [11].:

1. Classical Arabic which is the language of the holy Qur'an and seen in classical literature.
2. The Modern Standard Arabic that is understood by all Arabic-speakers, both are different mainly in the style and vocabulary.

In addition, these vowels are not see very often in regular Arabic text but it is important when handling the Holly Qur'an to avoid any reading mistakes that cloud change the word and its meaning [12].

There are number different features in Arabic Language that distinguish it from any other languages including [13]:

- Being a pro-drop language, where the subject can be encoded in the verb.
- Words in Arabic can be compound, where one word can be created from the morpheme and affixation as shown in Fig. 4 using the word "و بقولهم".

| word | مـــهـــولـــقـــب | | | و |
|---|---|---|---|---|
| components | مـــه | قَولـ | ـــب | و |
| translatitration | himm | qouli | bi | wa |
| translation | *their* | *say* | *by* | *and* |

**Fig. 4.** Compound Arabic word

- Gives the freedom of changing the order of the words within a sentence.
- Having a high word and letter ambiguity, where some Arabic words can be translated in to different number of English word, like the word "خالي" can be translated to three English words "empty", "imagined" or "battalion".

The characteristic of the Arabic language can produce many obstacles regarding the development of a new high intelligent technology and application such as Chatbots which Albalooshi et al. [14] mentioned in his study such as:

- The unique structure of the language can cause some difficulties when translation systems are trying to translate the English sentence into Arabic and vice versa.
- The richness of the morphology where one word such as the root "كتب" can generate number of different words "يكتب" and "he is writing", "تكتب" and "she is writing" which can be an obstacle for conversational agent.

Unlike the Latin, the Arabic language typographically is distinctive where it consist of 60 unique and different characters from letter to number, the writing direct which is basically start from the right to the left, the Arabic language is known with free word order meaning in addition to the regular order of the sentence (verb, subject and object) there is a predication sentence structure with no verbs, no diacritics that helps in making the pronunciation of the word, it is a clitic language where a full sentence can be presented by a single word, the inconsistent use of the punctuation marks and finally pro-drop issue.

### 2.3 Arabic Language and Chatbot Challenges

The Arabic language is facing many obstacles regarding the technology, application and the Internet which Albalooshi et al. [14] pointed about including the representation style of the language being unique and different that makes the task of the morphological analyzers in reading and understanding the sentence's meaning difficult.

Also, since the writing direction of the Arabic language is from the right to the left, many application and platform are still having problems in the representation and the direction of the words that could cause an issue in analyzing the Arabic text which is a major task in Chatbot applications and any other conversational systems [15].

Moreover, the difficulties in analyzing the sentence comes from the Arabic language is set to be a complex language system in the verb agreement, where the verbs agree with the gender [16]. In Arabic language, the verb "تقود" agree feminine and the verb "يقود" agree with masculine, where in the English the verb "drives" agree with both masculine and feminine.

# 3   Survey

This section will explore and review some of the most used and famous Chatbot development platforms and frameworks and see the level of support offered for the Arabic language in providing an easy and powerful working space for the Arabic Chatbot developer.

We will present brief survey on some of the existing Chatbot development frameworks and platforms including a summary comparison table of the level of support for the Arabic language and some desirable features to address the Arabic language and the Chatbot application challenges discussed above:

a. **Microsoft Bot Framework:**
It is a cross-platform and cloud based framework that uses two components which are the Bot Connector and Artificial Intelligent (AI), Natural Language Understanding (NLU) components to help in designing and deploying the Chatbot across range of services. The Microsoft Bot integration is impressive including Slack, Telegram, and Azure.

The Framework is known to be powerful and gives the developer the set of tools to edit the code along with provision of testing functionality. But on the other side it does not support the Arabic language and in order to use this framework, an English Chatbot is needed to be developed. Then, using a translation system the Arabic text must be translated first to English followed by process of fetching of the reply from the English Chatbot. This reply is then translated back to Arabic language to be presented in front of the Arabic user. Using this from of process can result in number of complexities and results in increase of error rate and cause many problems. For example, for the second person, English uses the word "You" whereas Arabic sees the gender of a person and then a specific pronoun is used. Such problems are evident in such kind of processes.

b. **Facebook Bot Engine (Wit.ai):**
This Framework is based on the Wit.ai technology that runs on a cloud server. The Engine is built to deploy a developed Chatbot in Facebook Messenger platforms. It uses Natural Language Processing (NLP), which means it can handle the typing mistakes. It includes Machine learning component where the developer feeds the Chatbot with a conversation sample to help it to understand and handle different forms and variation of the user's questions and inputs.

In addition, the Wit.ai technology allows the developers to extract the user's intent, sentiments from the user's input which helps in making the Chatbot more intelligent.

The Framework supports the Arabic Chatbot to be developed and deploy. But since the Arabic language is known to have a rich morphology and is a very diverse language, it is difficult and time consuming to train the Chatbot. Also, it requires some detailed knowledge about the Arabic language features to help in knowing what is the best way to extract the intent from the user's input.

c. **Dialogflow(Api.ai):**

It is a conversation platform to provide the Chatbot developers with tools to add Natural Language Processing (NLP) capabilities to the Chatbot and a built-in Machine learning (ML) to recognize the intent and context of what the users says which allows the Chatbot to provide a highly efficient and accurate responses.

In addition, it is powered by Google Cloud Speech which helps in expanding the Chatbot abilities to recognize voice interaction with a single API call. Also, deploying the Chatbot to a wide variety of platforms is possible with just a single click.

Dialogflow supports more than twenty different languages including English, French and Chinese. But till now the support for Arabic language is not included and the framework doesn't support developing an AI Arabic Chatbot.

d. **Pandorabots:**

It is a (AIML) based web service for building and deploying Chatbot and one of the largest and oldest hosting servers in the market. It allows the developers to create AI virtual agents to hold a human like conversation using text or voice with consumers. It's known to implement and support the AIML open standards and provide an API access to the Chatbot hosting platform. The developers can deploy the Chatbot in any application using its API.

Although it is multilingual and supports the Arabic language using easy and clear step by step building interface, some limitations can be seen when dealing with Arabic spelling mistakes and handling Informal Arabic language.

e. **IBM Watson Conversation:**

It is an AI platform for business from IBM and it's a collection of AI services designed to work together to create AI-powered conversational Chatbot with just a few lines of code. Also, using its speech to text and text to speech APIs helps in extending the chatbot capabilities to spoken conversation. Furthermore, there are many other services that can be added to the Chatbot such as Natural Language classifier to interpret and classify natural language with confidence, personality insights to enable deeper understanding of the user's personality.

Although the Arabic language is supported, there are some limitation when developing AI Arabic Chatbot. One its main features is the pre-defined sets of intent and entities including dates and location but unfortunately this feature is not yet available for Arabic language. In addition, since the Arabic writing direction is from the right to the left, there is still some problems in the conversation workspace toggling between right and left alignments. Finally, the developer can only work with Arabic through the conversation API.

f. **Chatfuel:**

It is one of the most-used, easy-to-create conversational chatbot platforms in the market. The idea behind it is having the ability to create a chatbot with some level of Artificial Intelligence (AI). It allows the developer to use the field of AI to give the Chatbot some human details. The platform is based on defining a set of keywords and phrases along with automatic responses.

It is a multi-language platform where the developer can choose from more than fifty languages available including the Arabic language. But, Chatfuel platform does not include Natural Language Processing (NLP) and the Arabic Chabot cannot be trained to recognize meaning or having the ability to analyze and process the Arabic language in its different forms to fully understand the language.

g. **Motion AI:**

This platform hand-holds the developer through developing every aspect of the Chatbot's flow. The developer can create a Chatbot to be used for Slack, Facebook Messenger or any other websites. It offers number of modules as a building block for the developers to choose from when creating their Chatbots and operates simply by designing a decision tree. Also, it provides a full report about the Chatbot post performance.

Few of its strong features are handling spelling mistakes, an excellent information and documentation and support of many languages including the Arabic Language. But because the developed Chatbot in this platform can only works on either a multiple choice or statement, the developer can't guide the conversation and

| Platform / Framework | Support for Arabic Language | Natural Language Processing feature | Artificial Intelligent feature | Arabic Entity Recognition feature | Handling of Informal Arabic | Handling of Verb Agreement | Control Conversation | Limitation |
|---|---|---|---|---|---|---|---|---|
| Microsoft Bot | NO | YES | YES | NO | NO | NO | NO | o Relying on a translation system help the Chatbot to understand the Arabic langue, but it increases the error rate and cause some problems in identifying the user's gender |
| Facebook Bot Engine (Wit.ai) | YES | YES | YES | NO | NO | NO | NO | o Time consuming when training the Chatbot to understand all the different forms of text in a highly-diversified language such as Arabic language |
| Dialogflow(Api.ai) | NO | YES | YES | NO | NO | NO | NO | o Limited language support. |
| Pandorabots | YES | YES | YES | NO | NO | NO | NO | o some limitations can be seen when dealing with Arabic spelling mistakes. |
| IBM Watson Conversation | YES | YES | YES | NO | NO | NO | NO | o The developer can only work with Arabic through the conversation API and not using the tooling interface. o Problems in the conversation workspace when toggling between right and left alignments of the Arabic language. |
| Chatfuel | YES | NO | YES | NO | NO | NO | NO | o the Chatbot cannot be trained to recognize meaning or typos. |
| Motion AI | YES | YES | YES | NO | NO | NO | NO | |

**Fig. 5.** Chatbot frameworks and platforms status with respect to Arabic Language

setting out specific question and responses which is not good for a much looser format conversations such as the Arabic language.

Figure 5 shows a table that summarize different issues related to Chatbot platforms and frameworks, including support for Arabic language along with some other desirable and essential features.

## 4 Conclusions

This paper reviewed the most used frameworks and platforms for Chatbot development and discussed their attitude towards Arabic language and Arabic Chatbot.

In conclusion, Arabic language is having less attention when it comes to creation of a Chatbot using development frameworks such as Microsoft Bot and Pandorabots. Although most of them support the Arabic language, some limitations were found when handling the Arabic language's special features such as the high morphology, informal Arabic language and text alignments. As the Arabic content is spreading faster over the Internet and has dominated the social media applications over the last few years, it is expected that more attention will be paid to provide support of the Arabic Language in the Chatbot applications in order to overcome the existing limitations and to provide better platforms to develop up-to-date and efficient Arabic Chatbots.

## References

1. Dohrmann, E., Hügi, J., Scheurer, N., Trummer, A., Schneider, R., Christine, E.: Bridging the virtual and the physical space: Kornelia - a chatbot for public libraries. In: BOBCATSSS (2010)
2. Kurdi, H., Alkhaider, S., Alfaifi, N.: Development and evaluation of a web based question answering system for Arabic language. Int. J. Natural Lang. Comput. 3(2), 12–32 (2014)
3. Woudenberg, A.: A Chatbot Dialogue Manager - Chatbots and Dialogue Systems: A Hybrid Approach, 1st edn. Open Universities Nederland (2014)
4. Jurafsky, D., Martin, J.: Speech and Language Processing (An Introduction to Natural Language Processing, Computational Linguistics and Speech Recognition), 1st edn. Library of Congress Cataloging-in-Publication Data, New Jersey (2000)
5. Amilon, M.: Chatbot with common-sense database. Royal Institute of Technology CSC School, Sweden (2015)
6. Chakrabarti, C., Luger, G.: A semantic architecture for artificial conversations. In: International Conference on Soft Computing and Intelligent Systems (SCIS) and 13th International Symposium on Advanced Intelligent Systems (ISIS) (2012)
7. Jusoh, S., Alfawareh, H.M.: Techniques, applications and challenging issue in text mining. IJCSI Int. J. Comput. Sci. Issues. 9(6), 431–436 (2012)
8. Abdul-Kader, S., Woods, J.: Survey on Chatbot design techniques in speech conversation systems. Int. J. Adv. Comput. Sci. Appl. 6(7), 72–80 (2015)
9. Sule, M.: Machine Language Techniques for Conversational Agents, 1st edn. University of North Texas, Texas (2003)
10. Elkateb, S., Black, W., Farwell, D., Pease, A., Vossen, P., Fellbaum, C.: Arabic WordNet and the challenges of Arabic. In: The Challenge of Arabic for NLP/MT Conference (2006)

11. Najeeb, M., Abdelkader, A., Zghoul, M.: Arabic natural language processing laboratory serving islamic sciences. Int. J. Adv. Comput. Sci. Appl. **5**(3), 114–117 (2014)
12. Muhtaseb, H., Mellish, C.: Some differences between Arabic and english: a step towards an Arabic upper model. In: The 6th International Conference on Multilingual Computing (1998)
13. Abuelyaman, E., Rahmatallah, L., Mukhtar, W., Elagabani, M.: Machine translation of Arabic language: challenges and keys. In: International Conference on Intelligent Systems, Modelling and Simulation (2014)
14. Albalooshi, N., Mohamed, N., Al-Jaroodi, J.: The challenges of Arabic language use on the internet. In: International Conference on Internet Technology and Secured (2011)
15. Attia, M.: Handling Arabic Morphological and Syntactic Ambiguity within the LFG Framework with a View to Machine Translation. School of Languages, Linguistics and Cultures (2008)
16. Alawneh, M., Omar, N., Sembok, T.: Machine translation from english to Arabic. In: International Conference on Biomedical Engineering and Technology (2011)

# The Generative Power of Arabic Morphology and Implications: A Case for Pattern Orientation in Arabic Corpus Annotation and a Proposed Pattern Ontology

Mohammed A. El-Affendi(✉)

Department of Computer Science, College of Computer and Information
Sciences, Prince Sultan University, Riyadh, Saudi Arabia
affendi@psu.edu.sa

**Abstract.** Most of current Arabic morphological analyzer use complex rules to handle the idiosyncrasies of certain Arabic word classes and special cases. The question that arises: is it feasible to design a pattern-oriented morphological analyzer that streamlines the process and avoid the use of complex rules? To answer this question a detailed study has been conducted using a small representative Arabic corpus. The study revealed that most of the words in the language can be generated using a limited number of patterns, morphemes and particles. Inflected and derivational words can be generated through combinations of roots and patterns. The total number of roots is around 10,000 while the total number of morphological patterns is below 1000. The total number of particles is around 325. Around 70% of words in the experimental corpus are templatic (based on morphological patterns). Although, the number of identified patterns reached 943, only a small subset of these is active. For example, the top 12 patterns in the identified list accounted for more than 50% of the generated templatic words. Although the total number of roots is around 10,000 the number of active roots is 3,461. Particles and similar morphemes account for around 30% of the text in the experimental corpus. These features greatly simplify the development of NLP applications such as spelling correctors, normalizers, lemmatizes and higher-level applications.

## 1 Introduction

Over the past few years, great efforts have been exerted to build corpus linguistic systems and provide resources for Arabic Natural Language Processing. Examples of these include the work of Khoja [10], Khoja et al. [11] who designed a tag set and built a part of speech tagger for Arabic.

Recently, significant achievements have been realized for Arabic NLP in three major international leading universities:

- The Stanford NLP group: achievements in Stanford University include the work of the NLP group of Stanford University (Toutanova et al. [1]), plus the work Diab et al. [4]

© Springer International Publishing AG, part of Springer Nature 2018
M. Alenezi and B. Qureshi (Eds.): *5th International Symposium*
*on Data Mining Applications*, pp. 36–45, 2018.
https://doi.org/10.1007/978-3-319-78753-4_4

- The Columbia NLP Group: The NLP of group of Columbia University have been generating very solid and significant contributions in the area of Arabic Corpus Linguistics, examples of which include the work of Habash and Rambow [8]. The work of Mona Diab with the Columbia group (Diab et al. [5]), the work of Habash et al. [6]. Habash also authored a very interesting book that summarizes many of the results and concepts related to Arabic NLP (Habash [7]). Not to forget the work of Roth et al. [9] on Arabic morphological tagging and diacritization.
- Leeds NLP Group: The NLP Group in Leeds University has also been giving considerable attention to Arabic Morphology, corpus tagging and other related areas. A good example of Leeds efforts is the seminal work of Dukes et al. [14] in annotating the Quran Corpus and the work of Sawalha and Atwell [2].

This paper, agrees with and supports the views of Habash and Rambow [8] regarding the complexity of Arabic morphology and the need for more flexible and wider tag sets. Interestingly, a similar view has been expressed in the work of the Leeds NLP group (Sawalha and Atwell [2]).

Accordingly, the paper proposes the design of a generalized Arabic morphological tag set based on morphological patterns and a corresponding pattern oriented Arabic morphological analyzer. A study has been conducted to investigate the significance and role of Arabic morphological patterns in generating Arabic text. In the study, we applied what we preach: a simple Arabic morphological tagger (ELAffendi and Altayeb 2014) [13] has been used to identify Arabic morphological patterns in a small experimental corpus and produce statistics regarding their frequency and generative power.

At the end, the paper proposes the design of a simple pattern ontology that links morphological patterns to part of speech categories and morphological features.

Research Question: In the context of building a comprehensive Arabic Morphological Toolkit (SAMT) (ELAffendi 2017) [15], based on the SWAM morphological analyzer (ELAffendi and Altayeb 2104) [13], an important question arose:

> "Is it feasible to build a pattern oriented templatic morphological segmentation algorithm that uniformly handle the segmentation process without the need for the complex rules used in current segmentation algorithms?"

This paper tries to answer the first part of the questions "the feasibility of building a pattern-oriented Arabic morphological analyzer". The second part of the question is addressed in a separate paper, where a data driven solution is used to handle the segmentation process uniformly without the need for the complex rules used by current analyzers (ELAffendi 2017) [15].

## 2 The Corpus

The study has been conducted using a small representative corpus consisting of the samples shown in Table 1 below. The samples have been collected from newspapers, books and free sources on the web. As clear from the table, the total number of words in the corpus is 489725. The total number of words that yielded to analysis is 443292. This study revealed that out of the 443292 words that yielded to analysis, 303606 are templatic words (68.4889%) and the remaining 139686 are nontemplatic (31.51106%).

**Table 1.** The experimental corpus used to conduct the study

| Category | Number of files | Total no. of words | Distinct words | Ratio T/Total |
|---|---|---|---|---|
| Religion | 9 | 63548 | 18733 | 12.97% |
| Social | 53 | 39809 | 9202 | 8.12% |
| Economics | 77 | 34728 | 5608 | 7.09% |
| Cultural | 110 | 125287 | 18341 | 24.48% |
| Political | 53 | 29097 | 3346 | 5.94% |
| Health | 17 | 9732 | 1372 | 1.98% |
| Technology | 72 | 45327 | 4083 | 9.25% |
| Sport | 92 | 44223 | 4625 | 9.03% |
| Art | 40 | 24724 | 2487 | 5% |
| Psychology | 5 | 8702 | 594 | 1.7% |
| Language | 2 | 4950 | 595 | 1% |
| Education | 19 | 35915 | 2307 | 7.3% |
| Philosophy | 2 | 2970 | 351 | 0.6% |
| Sciences | 15 | 20713 | 1860 | 4.2% |
| | **566** | **489725** | **73504** | **100%** |

## 3   The Frequency of Patterns

Table 2 below shows frequency of patterns whose frequency is higher than 0.50% (corresponding count > 2300). The results are very interesting and the findings may be summarized as follows:

- The number of distinct patterns that resulted from the analysis is 943. Most of these represent derivational forms, and in some cases inflectional forms crept in.
- The top 12 patterns in the list account for more than 50% of the templatic words in the corpus (51.319% to be exact). This means that more than 50% of the templatic words are generated by only 12 patterns (Fig. 2 and 3).

**Table 2.** The frequency of templatic words

| Pattern | Transliteration | Count | Relative frequency | Overall frequency |
|---|---|---|---|---|
| فعل | f3l | 56504 | 18.61096289 | 12.74645 |
| فعلة | f3lt | 13488 | 4.442599949 | 3.04269 |
| فعال | f3aal | 12899 | 4.248598513 | 2.90982 |
| يفعل | yf3l | 10361 | 3.412646654 | 2.337286 |
| فاعل | faa3l | 10087 | 3.322398108 | 2.275475 |
| تفعل | tf3l | 9211 | 3.033866261 | 2.077863 |
| تفعيل | tf3yl | 7792 | 2.566484193 | 1.757758 |
| فعيل | f3yl | 7635 | 2.514772435 | 1.722341 |
| مفعل | mf3l | 7611 | 2.506867453 | 1.716927 |

*(continued)*

**Table 2.** (*continued*)

| Pattern | Transliteration | Count | Relative frequency | Overall frequency |
|---|---|---|---|---|
| فعالة | f3aalt | 7224 | 2.379399617 | 1.629626 |
| فعول | f3wl | 6893 | 2.270376738 | 1.554957 |
| فعلية | f3ylt | 6102 | 2.009841703 | 1.376519 |
| أفعل | aa'f3l | 5034 | 1.658069999 | 1.135595 |
| فعيلة | f3ylt | 5029 | 1.656423127 | 1.134467 |
| مفعلة | mf3lt | 4684 | 1.542789009 | 1.05664 |
| فاعلة | faa3lt | 4384 | 1.443976733 | 0.988964 |
| فعلي | f3ly | 4268 | 1.405769319 | 0.962797 |
| أفعال | aa'f3aal | 4002 | 1.318155768 | 0.902791 |
| فعلت | f3lt | 3747 | 1.234165333 | 0.845267 |
| مفاعل | mfaa3l | 3381 | 1.113614355 | 0.762703 |
| فعله | f3lh | 3274 | 1.07837131 | 0.738565 |
| مفاعلة | mfaa3lt | 2880 | 0.948597854 | 0.649685 |
| افتعال | aaft3aal | 2815 | 0.927188527 | 0.635022 |
| فعلا | f3laa | 2589 | 0.852749946 | 0.584039 |
| فعلات | f3laat | 2585 | 0.851432449 | 0.583137 |
| إفعال | aa'f3aal | 2380 | 0.783910726 | 0.536892 |
| تفاعل | tfaa3l | 2309 | 0.760525154 | 0.520876 |

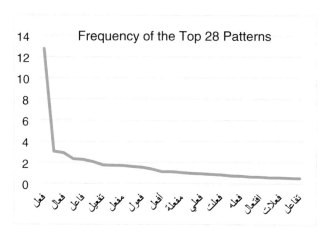

**Fig. 1.** Frequency of the top 28 patterns

- The top 53 patterns account for more than 80% of the templatic words in the corpus.
- Table 3 and Fig. 1 below provide a general idea about the frequency of patterns in relation to the number of words generated.
- It is interesting to note that more than 90% of the templatic words in the corpus has been generated by only 111 of the 943 patterns (11.77% of the patterns generate more than 90% of the words)

**Table 3.** Percent of words generated by the top n patterns

| Top n patterns | Per cent of templatic words generated |
|---|---|
| 2 | 23.05 |
| 4 | 30 |
| 8 | 40 |
| 12 | 51.3 |
| 18 | 60.344 |
| 29 | 70.295 |
| 53 | 80.219 |
| 111 | 90.048 |

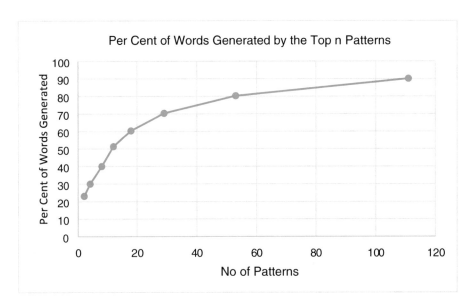

**Fig. 2.** Percent of words generated by the top n patterns in the list

- On the lower side of the table, the analysis revealed that 456 of the 943 patterns has a count of less than 10 in the corpus (see Table 4 below):

**Table 4.** Low frequency patterns

| Count in the corpus | Number of patterns |
|---|---|
| 1 | 132 |
| 2 | 108 |
| 3 | 50 |
| 4 | 48 |
| 5 | 30 |
| 6 | 24 |
| 7 | 14 |
| 8 | 21 |
| 9 | 16 |
| 10 | 13 |
| Total | 456 |

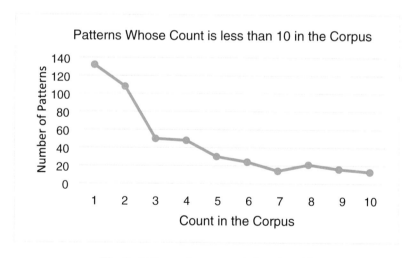

**Fig. 3.** Patterns whose count is less than 10

## 4 The Frequency of Nontemplatic Words (Particles and Foreign Nouns)

Table 5 below shows the summarized frequencies of nontemplatic words. The frequency has been computed relative to the total number of words that yielded to analysis (Fig. 4):

**Table 5.** Frequency of non-templatic words

| | Category | No of items per category | Category count | Category frequency % |
|---|---|---|---|---|
| 1 | Prepositions | 14 | 39122 | 8.825334 |
| 2 | Conditionals 1 | 10 | 16280 | 3.672523 |
| 3 | Temporal Circumstantial Adverbs | 60 | 14637 | 3.263946 |
| 4 | Relative Pronouns | 15 | 12335 | 2.78259 |
| 5 | Nasb (Accusative) Particles | 7 | 11535 | 2.60212 |
| 6 | Conjunction | 11 | 8483 | 1.913637 |
| 7 | Demonstrative Pronouns | 15 | 8191 | 1.826659 |
| 8 | Negation | 4 | 5908 | 1.332756 |
| 9 | Numerals | 95 | 5831 | 1.315391 |
| 10 | Pure Pronouns | 20 | 5095 | 1.149355 |
| 11 | Emphasis | 6 | 2090 | 0.471472 |
| 12 | Confirmation (Realization) | 1 | 1868 | 0.421393 |
| 13 | Exclusion | 7 | 1816 | 0.409663 |
| 14 | Locative Circumstantial Nouns | 20 | 1412 | 0.318528 |
| 15 | Foreign Nouns | 117 | 1263 | 0.284922 |
| 16 | Common | 12 | 1247 | 0.281305 |
| 17 | Conditionals 2 | 5 | 875 | 0.197388 |
| 18 | Frozen Verbs | 12 | 794 | 0.179116 |
| 19 | Jazm (Jussive) Particles | 2 | 46 | 0.010377 |
| 20 | Others | | 858 | 0.25258 |

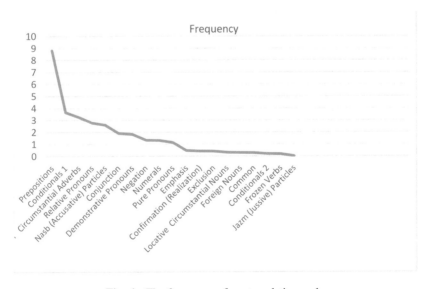

**Fig. 4.** The frequency of nontemplatic words

## 5  A Case for Pattern Orientation and a Proposed Ontology

It is clear from the above section that around 70% of Arabic text is generated by well-defined morphological patterns. The number of generating patterns is finite and as shown above 90% of the templatic text in the experimental corpus has been generated by only 111 patterns. The remaining 30% of the corpus text is accounted for by 316 nontemplatic particles and pronouns.

Arabic morphological patterns are well characterized and can easily be linked to morphological and part of speech features. Actually, some researchers developed full dictionaries for Arabic morphological patterns (Yagoub 1996) [12]. Accordingly, morphological patterns may serve as "feature centroids" for Arabic words. Identifying the exact morphological pattern for a given word automatically provides rich information regarding the associated morphological features.

It is expected that some ambiguity may arise in some cases (patterns may correspond to more than one-word class). Such cases may easily be resolved using hidden Markov models or any relevant disambiguating algorithm. Viterbi proved to be very effective in this respect.

Any context-aware morphological tagging algorithm may be used to identify the morphological patterns of templatic words. In our case, a simple two-phase algorithm has been used. In the first phase, the algorithm identifies the Surface Pattern (pattern without vocalization) using simple Bayesian inference. In the second phase, the algorithm applies Viterbi to identify the Deep Pattern (vowelized pattern).

The Surface Pattern in our case is the root of a weighted three level tree as shown in Fig. 5 below. As illustrated in the diagram a surface pattern may assume more than one deep pattern, depending on the context. The weights represent probabilities that lead to the appropriate deep pattern, again based on the context. Each deep pattern may correspond to more than one part of speech category. The weights on the branches are used to identify the appropriate part of speech category.

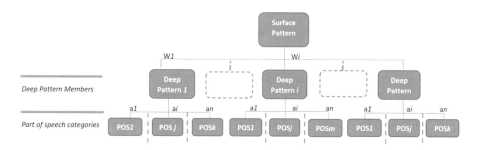

**Fig. 5.**  The surface pattern structure

## 6  Conclusion

A study has been conducted to investigate the role of morphological patterns in generating modern standard Arabic text. The main motivation is to provide support for the view that morphological patterns may be used as structural hubs for morphological features. The study revealed that around 70% of the corpus text represent templatic words that are generated from predefined morphological patterns. Not only that, but 90% of the corpus templatic words are generated by only 111 patterns (out of the listed 943 patterns). All morphological patterns can be characterized in detail and linked to morphological features and part of speech categories. A Hierarchical tree structure has been proposed to inform the tagging process. This structure has been practically used with the algorithm that generated the results for this study. It is recommended that this structure should be promoted to a full ontology. Clearly further work is needed to consolidate the findings and characterize the morphological patterns.

**Acknowledgement.** I would like to thank King Abdulaziz City for Science and Technology (KACST) for supporting this research work under NSTIP project 11-INF2159-04 "Arabic Spelling Checking and Correction".

# References

1. Toutanova, K., Klein, D., Manning, C.D., Singer, Y.: Feature-rich part-of-speech tagging with a cyclic dependency network. In: Proceedings of HLT-NAACL 2003, pp. 173–180 (2003)
2. Majdi, S., Atwell, E.: Fine-grain morphological analyzer & part-of-speech tagger for Arabic Text. In: Language Resources and Evaluation Conference, LREC (2010)
3. Diab, M.: Towards an Optimal POS tag set for Modern Standard Arabic Processing. In Proceedings of Recent Advances in Natural Language Processing (RANLP), Borovets, Bulgaria (2007)
4. Diab, M., Hacioglu, K., Jurafsky, D.: Automatic tagging of Arabic Text: from raw text to base phrase chunks. In: Proceedings of the 5th Meeting of the North American Chapter of the Association for Computational Linguistics/Human Language Technologies Conference (HLT-NAACL 2004), Boston, MA, pp. 149–152 (2004)
5. Diab, M., Ghoneim, M., Habash, N.: Arabic diacritization in the context of statistical machine translation. In: Proceedings of Machine Translation Summit (MT-Summit), Copenhagen, Denmark (2007)
6. Habash, N., Rambow, O., Roth, R.: MADA+TOKAN: a toolkit for Arabic tokenization, diacritization, morphological disambiguation, POS tagging, stemming and lemmatization. In: Proceedings of the 2nd International Conference on Arabic Language Resources and Tools (MEDAR), Cairo, Egypt (2009)
7. Habash, N.: Introduction to Arabic Natural Language Processing. Morgan & Claypool, San Rafael (2010)
8. Habash, N., Rambow, O.: Arabic tokenization, part-of-speech tagging and morphological disambiguation in one fell swoop. In: Proceedings of the 43rd Annual Meeting of the Association for Computational Linguistics (ACL 2005), Ann Arbor, Michigan, June 2005, pp. 573–580. Association for Computational Linguistics 2005
9. Roth, R., Rambow, O., Habash, N., Diab, M., Rudin, C.: Arabic morphological tagging, diacritization, and lemmatization using lexeme models and feature ranking. In: ACL-08: HLT, June 2008, Columbus, Ohio, pp. 117–120 (2008)
10. Khoja, S., Garside, R., Knowles, G.: A tagset for the morphosyntactic tagging of Arabic. In: Proceedings of Corpus Linguistics 2001, pp. 341–353, Lancaster, UK (2001)

11. Khoja, S.: APT: Arabic part-of-speech tagger. In: Proceedings of Student Research Workshop at NAACL 2001, Pittsburgh, pp. 20–26. Association for Computational Linguistics (2001)
12. Yagoub, A.B.: A Dictionary of Arabic Morphological Patterns (in Arabic). World of Books Publishing, Bierut (1996)
13. ELAffendi, M., Altayeb, M.: The SWAM Arabic morphological tagger: multilevel tagging and diacritization, using lexicon driven morphotactics and viterbi. In: ICAI 2014: The 2014 International Conference on Artificial Intelligence, 21–24 July 2014, Las Vegas, Nevada, USA (2014)
14. Dukes, K., Atwell, E., Sharaf, A.B.M.: Syntactic annotation guidelines for the Quranic Arabic dependency treebank. In Proceedings of the Language Resources and Evaluation Conference (LREC) (18221827), Valletta, Malta (2010b)
15. ELAffendi, M.A., Abuhaimed, I.: SWAM Arabic morphological toolkit: a hybrid neuro model for segmentation. POS Tagging and Spellchecking (in press)

# A Benchmark Collection for Mapping Program Educational Objectives to ABET Student Outcomes: Accreditation

Addin Osman[1(✉)], Anwar Ali Yahya[1,2], and Mohammed Basit Kamal[1]

[1] College of Computer Science and Information Systems,
Najran University, Najran, Saudi Arabia
{aomaddin,aaesmail,mbkamal}@nu.edu.sa
[2] Faculty of Computer Science and Information Systems,
Thamar University, Thamar, Yemen

**Abstract.** This research aims to present a collection of dataset, which represents the mapping of program education objectives to the ABET student outcomes. The dataset has been collected by the authors from 32 self-study reports from Engineering programs accredited by ABET, which are available online. The paper presents the constraints under which, the dataset was produced, because its understanding plays a vital role in the usage of this collection in future researches. To illustrate the properties and usefulness of the collection, the dataset has been cleansed, preprocessed, some features have been selected, then it has been benchmarked using nine of the widely used supervised multiclass classification techniques (Binary Relevance, Label Powerset, Classifier Chains, Pruned Sets, Random k-label sets, Ensemble of Classifier Chains, Ensemble of Pruned Sets, Multi-Label k Nearest Neighbors and Back-Propagation Multi-Label Learning). The techniques have been compared to each other using five well-known measurements (Accuracy, Hamming Loss, Micro-F, Macro-F, and Macro-F). The Ensemble of Classifier Chains and Ensemble of Pruned Sets have achieved encouraging performance compared to the other experimented multi-label classification methods. The Classifier Chains method has shown the worst performance. In general, promising results have been achieved. New research directions and baseline experimental results for future studies in educational data mining in general and in accreditation in specific have been provided.

**Keywords:** Benchmark collection · Program educational objectives
Student outcomes · ABET · Accreditation · Machine learning
Supervised multiclass classification · Text mining

## 1 Introduction

Supervised text classification or categorization is the process of automatically and correctly assigning a given natural language corpus into one or more predefined class-label(s) [1–4]. In the basic classification processes, each input is considered

© Springer International Publishing AG, part of Springer Nature 2018
M. Alenezi and B. Qureshi (Eds.): *5th International Symposium*
*on Data Mining Applications*, pp. 46–60, 2018.
https://doi.org/10.1007/978-3-319-78753-4_5

independently from all other inputs and the class-labels are specified in advance. In machine learning, Multi-Label Classification (MLC) or multinomial classification is the process of classifying instances into more than two classes. Classifying instances into one of two classes is known as binary classification. Supervised multiclass classification techniques aim at assigning a different number of class-labels for each instances of a dataset. Given a training data set of the form $(x_i, y_i)$, where $x_i \in \Theta$ is the $i^{th}$ example of a dataset $\Theta$ and $y_i \in \{1,..., k\}$ is the $i^{th}$ class-label of $k$ classes. We aim at finding a learning model $\Phi$ such that $\Phi(x_i) = y_i$ for new unseen examples [5–7]. A supervised classifier is a classifier, which is constructed on training corpora including the correct class-labels for each input and its framework is shown in Fig. 1.

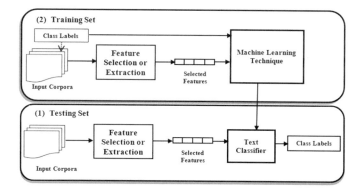

**Fig. 1.**   A framework for supervised text classifier

Figure 1 shows a supervised text classifier framework, in which:

- The first part of the framework shows that during training, a feature selection or extraction is used to convert each input corpus value to a feature set. These feature sets capture the basic information about each input corpus. Feature sets together with their class-labels are fed into machine learning technique(s) to generate a text classifier.
- The second part of the framework shows that during testing, the same feature selection or extraction is used to convert unseen input corpora to feature sets. These feature sets are then fed into the text classifier, which generates the predicted class-labels.

Supervised text classification has been tremendously investigated in many research fields such as machine learning, information retrieval and computational linguistics [1, 8–10]. It is widely applied in categorizing newspaper articles and news wire contents into topics (e.g. sports, technology and politics), categorizing web pages into news/media, social network, education, filtering emails into spam or not, and sentiment analysis [11, 12]. These are corpora to which, human indexers have assigned categories from predefined sets. Text corpora enable researchers to test ideas without hiring indexers, and (ideally) to objectively compare results with published studies. However, the investigation of text classification in academic accreditation, which is a very rich area of a diverse of documents and collections is very limited. Accreditation is an internationally recognized evaluation process used to assess and improve the quality, efficiency, and effectiveness of educational institutions and programs. It is widely imposed in most of the educational institutions and programs worldwide.

Accreditation Board for Engineering and Technology (ABET) paved the way for the accreditation process in the USA since 1932. It has been widely accepted and recognized as a leading accreditor of colleges and university programs in applied science, computing, engineering and technology in the USA and worldwide. Through its accreditation activities and dedication, ABET has accredited more than 2,700 programs at over 550 colleges and universities nationwide [13].

ABET delegates the responsibility for evaluating and taking accreditation actions on baccalaureate programs in engineering to the Engineering Accreditation Commission (EAC). The EAC depends on experts, volunteers, visiting teams to perform the crucially important and fundamental evaluation that is the basis of the EAC/ABET accreditation process. The engineering programs that applying for ABET accreditation must submit to ABET their Self-Study Reports (SSRs). SSRs are the primary documents, which programs use to demonstrate their compliance with all applicable ABET criteria and policies. The SSR is the foundation for the review team's judgment of whether the program meets ABET criteria for accreditation. It addresses all paths to completion of the degree, all methods of instructional delivery used for the program, and all remote location offerings [13, 14].

All programs applying for ABET accreditation must have documented program educational objectives (PEOs) that are consistent with the mission of the institutions and the needs of the programs' various constituencies. ABET defines PEOs as the expected accomplishments of graduates during the first few years after graduation. There must be a documented and effective process for the periodic review and revision of the PEOs. Related to that, the programs must have documented student outcomes (SO) in its SSR which, describe what students are expected to know and be able to do by the time of graduation [13–15]. The PEOs are manually mapped to the ABET SOs which, represent one of the most important parts of the SSRs.

This research is an endeavor to provide a dataset collection of mapping PEOs to SOs and to shed light on a rich new research area (accreditation) for the researchers.

The rest of the paper is organized as follows. In Sect. 2 a literature review has been provided, which summarizes some of the significant works related to EDM. In Sect. 3, the importance of the data set has been presented. In Sect. 4, the collection has been presented, containing the operational setting under which, the collection has been produced, the content of the collection, and the quality of the collection. In Sect. 5, the benchmarking of the collection to different multi-class classification has been presented. In Sect. 6, the results and discussions of the benchmark have been presented. Lastly, in Sect. 7, a summary of the work and most of the significant future directions of the research has been presented.

## 2    Literature Review

There is a rapid growth in the amount of published data collections in all research areas. The collections are deluging all research areas such as e-commerce, bioinformatics, social networks, and recently, the area of education, which is generally called Educational Data Mining (EDM) [16].

There is a review of 81 works (from 1995 up to 2005) in EDM [17]. The works were divided into three distinct groups of eLearning systems: Web-Based Courses (WBC), Learning Management Systems (LMS), and Adaptive and Intelligent Web-Based Educational Systems (AIWBES). However, from all these works, only 7 of them were performed for unstructured data (4 of them related to WBC, 2 of them to LMS, 1 to AIWBES).

Baker and Yacef presented a review summarized the works of 45 EDM references, where one published in 1973, another in 1995, and one more in 1999 [18]. This review specified some EDM targets, such as: student models, models of domain knowledge, pedagogical support, and impacts on learning. Another review provided 91 references about three topics: computer based education systems, DM, and EDM. The EDM works have been organized into four functionalities: student modeling, tutoring, content, and assessment [19].

Romero and Ventura augmented their previous EDM review by including new 225 works, and three more papers published in 2010. One novelty concerns a list of 235 works classified and counted in the following way: 36 in traditional education, 54 WBES, 29 LMS, 31 ITS, 26 adaptive educational systems, 23 test-questionnaires, 14 text-contents, and 22 others [20].

A report presented by Ihantola et al. provides a survey of the body of knowledge concerning the use of EDM and Learning Analytics (LA), which focused on teaching and learning of programming between years 2005 and 2015. They reported a significant growth in research work in the field [21].

In a paper published by Fatima et al. a wide variety of applications of EDM have been discussed, i.e. improving student models, discovering or improving models of the knowledge structure of the domain, studying the pedagogical support provided by learning software, scientific discovery about learning and learners [22].

A systematic literature review has been presented by Papamitsiou and Economides [23], which provided a comprehensive background for understanding current knowledge on LA and EDM and its impact on adaptive learning. They examined the literature on experimental case studies conducted in the domain during six years between 2008 and 2013. Another research works have been presented by Isha et al. and Raheela et al. [24, 25] which, used DM techniques to study students' performance.

The literature shows few studies in EDM related to the area of quality assurance and accreditation. Some of these studies have utilized the Bloom's taxonomy to automatically classify questions. The educational institutions are required to verify the correctness of the classification of exam questions according to Bloom's Cognitive Levels. The educators and quality community are facing a tremendous number of questions to check and evaluate manually [26–28].

Currently, most of the researches on EDM pay great attention towards the use of e-learning like Moodle, WebCT, Blackboard and some are using their private tools for the learning purposes. With regard to future research, the research in EDM area can shift from the e-Learning to the quality assurance and accreditation. The SSRs can be a treasure for unstructured data to be utilized by researchers for many activities in the EDM area. It is important to notice that the traditional educational data sets are normally small (short text) if we compare them to databases used in other data mining fields such as ecommerce applications that involve thousands of clients [29, 30]. One of the greatest and missed value of sharing a data collection is providing excellent interpretation and analysis of the collection that presents its quality and suitability for use for specific purpose.

## 3  Importance of the Data

In [30], it has been mentioned that much of the data collections represent the base for the scientific research and they are changing the face of the world. Scientific data collections are, at least, intermediate results in many scientific research projects. An interesting situation is that data collections are becoming more significant themselves and can sometimes be considered as the primary intellectual output of the research.

Data collections do not exist in a vacuum. The quality and usefulness of the collections depend on the context on how they are collected, processed, analyzed, validated, and interpreted. Thereafter, can be used to lead to a scientific publication. The process leading to a scientific publication can be described by the following basic scientific workflow shown in Fig. 2 [31]:

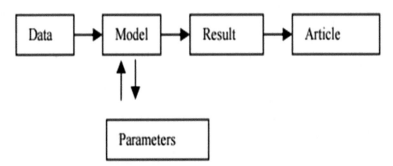

**Fig. 2.** The process leading to scientific publication [31]

The research interests in text categorization has been tremendously increasing in machine learning, information retrieval, natural language processing, computational linguistic, and data mining. Text categorization is a significant application area for machine learning and mainly depend on the availability of the collections of text categorization [1, 8, 9, 11]. Generally, the academic programs planning for accreditation or reaccreditation prepare their text categorization collections depending on human experts to assign class-labels from a predefined set. The accreditation bodies also depend on human experts to check, read and comprehend these data collections, specifically in the case of PEOs. At the beginning, they have to check the satisfaction of the PEOs to the specification of objectives that mentioned before, then to check the correctness of the mapping to the appropriate SOs. It is inevitable that a large amount of manual human effort and resources are required [32–34].

The text collections in accreditation are suffering from many weaknesses such as lack of complete text, peculiar textual properties, scattered text collections, and/or limited availability of corpora data. These difficulties are worsened by the shortage of documentation providing the methods used in producing these collections and on the nature of their classification systems. The researchers interested in text classification were facing serious problems, because of the shortage of well prepared collections and documentation. They have often been compelled to impose their own classes [11, 35, 36].

## 4 Data Collection

In this research, we have considered the data of the SSRs were produced in an operational settings at ABET accredited engineering programs under procedures managed and operated by specialists and experts in the area. The data were seldom used in research, only later the use of the data in research is contemplated.

We have sent requests to most of the ABET accredited programs in different countries requesting their SSRs. Nevertheless, we have not received any response from any program. At the end we have collected the data from SSRs of 32 ABET accreted Engineering programs, which are available online. The high frequency of the online publication of SSRs in Engineering encouraged the researchers to select them for this research.

### 4.1 Documents

The SSRs are one of the largest international source of accreditation data. Hundreds of programs in engineering have been accredited by ABET worldwide. The SSRs are produced in English language and many of them distributed and made available online in PDF and word documents.

The data of mapping PEOs to ABET SOs were drawn from those SSRs and only the ones that are in English language were considered for this research. The data collection has been formatted in XML.1, Excel and ARFF formats. The preparation of the dataset involved substantial verification and validation of the content, attempts to remove spurious or duplicated objectives, fulfilling the objectives and outcomes format, etc.

### 4.2 Class-Labels (Student Outcomes)

All ABET accredited programs must have documented PEOs, which are compatible with the institution's mission and the requirements of the diverse connstituencies of the program. ABET defines PEOs as the expected accomplishments of graduates during the first few years after graduation. The periodic review and revision of the PEOs must be documented. In addition to that SOs must clearly present what students will know and be able to do by the time of graduation. These are related to the knowledge, skills, and behaviors that students develop while they advance in the program. The PEOs were manually mapped to the eleven ABET SOs, which are considered as the class labels. One PEO can be mapped to more than one SO (supervised multiclass classification). SOs are outcomes (a) through (k) plus any additional outcomes that may be articulated by the program [14, 15]. In our case, we have considered only the ABET SOs, which are [11, 15]:

a. *an ability to apply knowledge of mathematics, science, and engineering*
b. *an ability to design and conduct experiments, as well as to analyze and interpret data*
c. *an ability to design a system, component, or process to meet desired needs within realistic constraints such as economic, environmental, social, political, ethical, health and safety, manufacturability, and sustainability*
d. *an ability to function on multidisciplinary teams*
e. *an ability to identify, formulate, and solve engineering problems*
f. *an understanding of professional and ethical responsibility*
g. *an ability to communicate effectively*
h. *the broad education necessary to understand the impact of engineering solutions in a global, economic, environmental, and societal context*
i. *a recognition of the need for, and an ability to engage in life-long learning*
j. *a knowledge of contemporary issues*
k. *an ability to use the techniques, skills, and modern engineering tools necessary for engineering practice.*

### 4.3   Policies of Developing PEOs

Developing POEs policies specify certain requirements to be satisfied while developing PEOs. In this section, some of the well-known policies employed in developing PEOs are presented. The PEOs of a program should clearly specify the intent of the program and what it is expected to achieve. There should clearly be no any room for misinterpretation and misunderstanding. Every PEO must start with a relevant action verb selected from the Bloom's Taxonomy that specify and definite observable and required behaviors [17–19].

POEs should also specify or state what learners are expected to do or be able to do upon completion of the program. A very important point to be noted is that each and every learner completes the program should be able to do something new and better than before. In addition, the PEOs should be **SMART**:

- **S**pecific: clear about what, where, when, and how the situation will be changed.
- **M**easurable: able to quantify the targets and benefits.
- **A**chievable: able to attain the objectives (knowing the resources and capacities at the disposal of the community).
- **R**ealistic: able to obtain the level of change reflected in the objective.
- **T**ime bound: stating the time period in which they will each be accomplished.

## 4.4   Data Cleansing and Statistics

The data used in this research has been passed through different cleansing processes so as to be transformed into a usable form. Some libraries of Natural Language Processing in Python have been utilized for the data cleansing and calculating some statistics of the data [37]. At the beginning, the vocabularies of the corpus of all *PEOs* have been computed, then all items that occur in an existing wordlist have been removed. Only the uncommon or misspelled words have been left. As a result, no uncommon or misspelled words have been found.

Some statistics have been performed on the data set. The number of all words (tokens) is *8118*, the average length of the tokens in the *PEOs* corpus is *6.20* and the most *50* frequent words have been derived from the corpus. They are shown below together with their frequencies:

*(and, 591), (to, 340), (in, 307), (the, 256), (of, 245), (engineering, 194), (will, 132), (professional, 117), (their, 112), (a, 109), (or, 102), (graduates, 94), (skills, 81), (be, 65), (ability, 64), (as,62), (an, 59), (with, 56), (design, 53), (problems, 52), (for, 50), (students, 49), (knowledge, 49), (learning, 48), (effectively, 47), (technical, 45), (systems, 45), (our, 43), (work, 43), (graduate, 42), (development, 42), (apply, 41), (have, 39), (through, 38), (demonstrate, 38), (ethical, 35), (leadership, 33), (program, 32), (environmental, 31), (pursue, 30), (practice, 28), (education, 28), (career, 27), (technology, 27), (life, 27), (lifelong, 26), ((, 26), ((, 26), (on, 25), (communicate, 25).*

Figure 3 shows the cumulative frequency plot for *50* most frequently words in the corpus, which account for nearly half of the tokens.

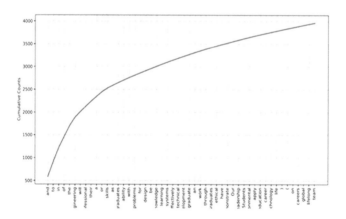

**Fig. 3.** Cumulative frequency plot for 50 most frequently words in the Corpus: these account for nearly half of the tokens

The tokens of length greater than or equal to *15* are:

*professionalism, competitiveness, multidisciplinary, collaboratively, instrumentation, internationally, microprocessors, interdisciplinary, telecommunications, responsibilities, electromechanical, interrelationships, entrepreneurial, accomplishments, microcontrollers.*

Here are all tokens from the corpus that are longer than seven characters, which also occur more than seven times (87 words were found):

*Graduates, Students, activities, advanced, analysis, applying, appreciation, appropriate, awareness, changing, commitment, communicate, communication, community, components, computer, continue, continuing, continuous, critical, demonstrate, development, disciplinary, disciplines, economic, education, effective, effectively, electrical, electronic, engineering, engineers, environment, environmental, experiments, function, fundamental, graduate, graduates, identify, improvement, including, industrial, industry, information, knowledge, leadership, learning, licensure, lifelong, management, mathematics, mechanical, multidisciplinary, necessary, organizations, participate, personal, practical, practice, practices, prepared, principles, problems, processes, productive, profession, professional, professionally, programs, research, resources, responsibilities, responsibility, societal, software, solutions, students, successful, successfully, teamwork, technical, techniques, technology, thinking, training, understanding.*

The process of searching for words with long lengths helped in detecting and fixing many two concatenated words (e.g. "*willdemonstrate*" and "*lifelonglearning*"). The concatenation of words resulted from combining *PEOs* from different sources and transforming them from *Excel* format to text format.

Table 1 illustrates some of the most important statistics and specification of the collection.

**Table 1.** The table summarizes important statistics of the collection

| No. of Instances | No. of Attributes | No. of Labels | Label Cardin. | Validation Method | No. Training Instances | No. Testing Instances |
|---|---|---|---|---|---|---|
| 167 | 160 | 11 | 3.8 | Training/ Testing split (66%, 34%) | 110 | 57 |

## 5    Benchmarking of the Collection

The aim behind the benchmarking of the data collection is to make sure that the collection is accurate and valid. The collection has been divided into training and testing sets to measure the effectiveness of some machine learning on the collection. Some multi-label classification methods and supervised learning algorithms are used in this work. The collection has been preprocessed to convert the data instances into a representation suitable for machine learning algorithms.

### a.    Training and Testing Sets

The collection has been preprocessed to convert the data instances into a representation suitable for machine learning algorithms. Table 2 presents the specifications of the main preprocessing steps of the collection.

**Table 2.** The specification of the main processing steps implemented on the collection

| Step | Specification |
|------|---------------|
| Tokenization | Alphabetic tokenize |
| Lowercase tokens | True |
| Stop words handler | Rainbow |
| Stemmer | Snowball stemmer |
| Min. Term frequency | 3 |
| TF transform | True |
| IDF transform | True |
| Normalization | All data |

b. **Measuring the Effectiveness of Techniques**

In the experiments, we used various evaluation measures that have been suggested by Tsoumakas et al. [38]. More specifically, to evaluate different MLC methods, the following evaluation measures were used: Accuracy (Acc.), Hamming Loss (HL), Micro-F, Macro-F (Example-based), Macro-F (Label-based).

c. **Machine Learning Techniques**

All MLC methods and supervised learning algorithms used in this work are implementations of the Weka-based [35] package of Java classes for MLC, called Meka [39]. This package includes implementations of some of MLC methods most widely applied in the literature. All the algorithms were supplied with Weka's J48 implementation of a C4.5 tree classifier as a single-label base learner.

In the experiment, nine MLC methods have been used, where:

- Four of them are problem transformation methods: Binary Relevance (BR), Label Powerset (LP), Classifier Chains (CC), Pruned Sets (PS)
- Three of them are ensemble methods: Random k-label sets (RAkEL), Ensemble of Classifier Chains (ECC) and Ensemble of Pruned Sets (EPS)
- And the remaining two are algorithms adaptation methods: Multi-Label k Nearest Neighbours (ML-kNN) and Back-Propagation Multi-Label Learning (BPMLL).

d. **Parameter Configuration**

All configurable parameters of the participating algorithms were set to their optimal values as reported in the relevant papers. For BR, LP and CC no parameters were required. The PS methods required two parameters p and strategy parameter for each dataset. We used $p = 1$ and strategy parameters ($A_b$) for all datasets with value of $b = 2$ as proposed by [40]. Whereas for the ensemble methods configuration, the number of models in the ECC methods was set to 10 as proposed by [41]. For RAkEL the number of models was set to 10 and the size of the label-sets (K) was set to half the number of labels. For EPS, at each dataset p and strategy, the parameters were set to the same values as those used for the PS method. EPS requires an additional parameter (the number of models) which was set to 10. For all ensemble methods the majority voting threshold was set to 0.5.

## 6    Results and Discussion

Table 3 summarizes the results obtained using the above-mentioned evaluation measures.

**Table 3.** Performance of MLC methods

|        | Acc.  | HL    | Micro-F1 | Macro-F1 (by example) | Macro-F1 (by label) |
|--------|-------|-------|----------|-----------------------|---------------------|
| BR     | 0.274 | 0.391 | 0.442    | 0.389                 | 0.427               |
| LP     | 0.259 | 0.34  | 0.42     | 0.352                 | 0.411               |
| CC     | 0.264 | 0.325 | 0.37     | 0.335                 | 0.363               |
| PS     | 0.284 | 0.324 | 0.425    | 0.366                 | 0.41                |
| RAkEL  | 0.338 | 0.365 | 0.501    | 0.462                 | 0.469               |
| ECC    | 0.358 | 0.357 | **0.547**| 0.48                  | **0.534**           |
| EPS    | **0.368** | **0.397** | 0.518 | **0.49**           | 0.515               |
| BPNN   | 0.281 | 0.37  | 0.37     | 0.377                 | 0.456               |
| MLkNN  | 0.32  | 0.351 | 0.5      | 0.45                  | 0.481               |

The results of the experimental comparison revealed that the EPS and ECC perform better than the remaining methods. More specifically the ECC outperforms the remaining methods in terms of Micro-F1 and Macro-F1 (by label), whereas the EPS outperforms them in terms of accuracy, HL, and Macro-F1 (By Example). It is also obvious that the CC perform the lowest. It should be mentioned that the performances of these methods vary based on the properties of the data set. For example, it has been pointed out in [38] that the label cardinality and label density are the key parameters that influence the performance of MLC methods. More specifically, the low values of the label cardinality and label density justify the better performance of BR technique compared to other techniques in [42]. In our work, the small size of the data set and the high value of label cardinality are the key factors that influence the performance of the experimented methods.

On the other hands, Table 4 shows a different view of the obtained results in terms of the accuracy per label. As it can be noted, the PS and CC methods show a comparable performance, which remarkably outperform the remaining methods, whereas the BR and EPS show a comparable and lowest performance. Again, the properties of the data set, small size and high value of label cardinality, plays a key role in determining the performances of these methods.

**Table 4.** Per label accuracy of MLC methods

| MLC method | Label | | | | | | | | | | | |
|---|---|---|---|---|---|---|---|---|---|---|---|---|
| | SO-a | SO-b | SO-c | SO-d | SO-e | SO-f | SO-g | SO-h | SO-i | SO-j | SO-k | Average Acc. |
| BR | **0.754** | 0.632 | 0.684 | 0.351 | 0.667 | 0.702 | **0.667** | 0.719 | 0.667 | 0.298 | 0.561 | 0.609 |
| LP | 0.719 | 0.561 | **0.719** | 0.719 | 0.667 | 0.596 | **0.667** | 0.667 | **0.719** | 0.614 | 0.614 | 0.66 |
| CC | 0.719 | 0.702 | 0.684 | **0.754** | **0.702** | 0.684 | 0.632 | 0.614 | 0.684 | 0.649 | 0.596 | **0.675** |
| PS | 0.702 | 0.596 | 0.684 | **0.754** | 0.684 | **0.737** | 0.579 | 0.719 | 0.632 | **0.737** | 0.614 | **0.676** |
| RAkEL | 0.719 | **0.719** | 0.579 | 0.649 | 0.667 | 0.544 | 0.579 | 0.684 | 0.702 | 0.579 | 0.561 | 0.635 |
| ECC | 0.737 | 0.632 | 0.614 | 0.667 | 0.667 | 0.596 | 0.596 | 0.684 | 0.614 | 0.632 | **0.632** | 0.643 |
| EPS | 0.614 | 0.667 | 0.491 | 0.561 | 0.544 | 0.544 | 0.544 | **0.772** | 0.649 | 0.632 | 0.614 | 0.603 |
| BPNN | 0.719 | 0.579 | 0.526 | 0.684 | 0.632 | 0.509 | 0.649 | 0.719 | 0.667 | 0.684 | 0.561 | 0.63 |
| MLkNN | 0.719 | 0.702 | 0.632 | 0.719 | 0.649 | 0.544 | 0.579 | 0.702 | 0.649 | 0.649 | 0.596 | 0.649 |

Finally it should be mentioned that due to the small size of the data set, the above mentioned results cannot be considered as conclusive. However, these results give an initial impression on the MLC in the context of EDM.

# 7 Conclusion and Future Work

The available data collections are the driving force, which propelling the research in unstructured text classification in general and in EDM in specific. The successful benchmark conducted on the collection (mapping of program educational objectives to ABET student outcomes) presented in this research has the potential to substantially contribute to the advancement of the research in unstructured text classification in accreditation processes and to shed light on different aspects of accreditation processes as a new research area related to educational data mining, text mining and machine learning.

The experimental results show that the EPS and ECC significantly performs better than do the other seven MLC methods (BR, LP, CC, PS, RAkEL, ML-kNN, and BPMLL) in terms of Micro-F1 and Macro-F1 (by label), whereas the EPS outperforms them in terms of accuracy, Hamming Loss, and Macro-F1 (by example). It is also obvious that the worst performance is accomplished by the CC method. It should be noted that the performance of these methods vary based on the properties of the data set and the model assumed to fit with the model. Although the EPS and ECC have achieved encouraging performance compared to the other experimented MLC methods, much work remains for further study. Some interesting and important research problems include:

- The data collection can be extended and generalized by collecting more data and performing further investigations on the extended collection.
- The approach/philosophy of exploratory data analysis can be implemented on the data collection to get insight into it, and detect outliers and anomalies.
- The approach/philosophy of exploratory data analysis can be employed to uncover the underlying assumptions behind the collection and uncover its structure. Testing of underlying assumptions is a tool for the validation and quality of the selection of

an appropriate model that best fit with the collection. In addition, it extracts a good-fitting (parsimonious) models. The models can be classification, clustering, association rules, etc.

- Since the lengths of the objectives are short, some techniques of short text understanding can be implemented on the data set to enhance the classification accuracy. These techniques can be short text similarity, short text classification, mapping text to semantic spaces (embedding), and using deep learning to model latent semantic representation for short texts.

# References

1. Fabrizio, S.: Machine learning in automated text categorization. ACM Comput. Surv. **34**(1), 1–47 (2002)
2. Shweta, C.D., Maya, I., Parag, K.: Empirical studies on machine learning based text classification algorithms. Adv. Comput. Int. J. (ACIJ) **2**(6), 161–169 (2011)
3. Fabricio, A.B., Daniel, C., Guimarães P.: Combined unsupervised and semi-supervised learning for data classification. In: IEEE 26th International Workshop on Machine Learning for Signal Processing (MLSP), Salerno, Italy, pp. 13–16 (2016)
4. Lunke, F., Yong, X., Xiaozhao, F., Jian, Y.: Low rank representation with adaptive distance penalty for semi-supervised subspace classification. Pattern Recogn. **67**, 252–262 (2017). http://ieeexplore.ieee.org/xpl/mostRecentIssue.jsp?punumber=7605057
5. Bishop, C.M.: Pattern Recognition and Machine Learning. Springer, Secaucus (2006)
6. Murphy, K.P.: Machine Learning: A Probabilistic Perspective, 1st edn. The MIT Press, Cambridge (2012)
7. Duda, R.O., Hart, P., Stork, D.: Pattern Classification, 2nd edn. Wiley-Interscience, New York (2000)
8. David, D.L., Robert, E.S., James, P.C., Ron, P.: Training algorithms for linear text classifiers. In: Proceedings of the 19th Annual International ACM SIGIR Conference on Research and Development in Information Retrieval (SIGIR 1996), pp. 298–306. ACM, New York (1996)
9. David, D.L.: Reuters-21578 text Categorization test collection. Distribution 1.0. Readme file (version 1.2). Manuscript (1997)
10. Yiming, Y.: An evaluation of statistical approaches to text categorization. Inf. Retrieval **1**(1–2), 67–88 (1999)
11. David, D.L., Yiming, Y., Tony, G.R., Fan, L.: RCV1: a new benchmark collection for text categorization research. J. Mach. Learn. Res. **5**, 361–397 (2004)
12. Pratiksha, Y., Gawande, S.H.: A comparative study on different types of approaches to text categorization. Int. J. Mach. Learn. Comput. **2**(4), 423–426 (2012)
13. ABET, ABET Strategic Plan, Accreditation Board for Engineering and Technology, Inc., ABET, 1 November 1997
14. Engineering Accreditation Commission (ABET), Criteria for Accrediting Engineering Programs Effective for Review During the 2015–2016 Accreditation Cycle, 415 N. Charles Street Baltimore, MD 21201, United States of Ameriaca, ABET (2014)
15. ABET, Criteria for Accrediting Engineering Programs Effective for Reviews During the 2016–2017 Accrediting Cycle
16. de Baker, R.S.J.: Data mining for education. In: McGaw, B., Peterson, P., Baker, E. (eds.) International Encyclopedia of Education, 3rd edn. Elsevier, Oxford (2010)
17. Romero, C., Ventura, S.: Educational data mining: a survey from 1995 to 2005. Expert Syst. Appl. **33**(1), 135–146 (2007)

18. de Baker, R.S.J., Yacef, K.: The state of educational data mining in 2009: a review and future vision. J. Educ. Data Min. **1**(1), 1–15 (2009)
19. Peña-Ayala, A., Domínguez, R., Medel, J.: Educational data mining: a sample of review and study case. World J. Educ. Technol. **2**, 118–139 (2009)
20. Romero, C., Ventura, S.: Educational data mining: a review of the state of the art. IEEE Trans. Syst. Man Cybern. Part C Appl. Rev. **40**(6), 601–618 (2010)
21. Ihantola, P., Vihavainen, A., Ahadi, A., Butler, M., Börstler, J., Edwards, S.H., Isohanni, E., Korhonen, A., Petersen, A., Rivers, K., Rubio, M.Á., Sheard, J., Skupas, B., Spacco, J., Szabo, C., Toll, D.: Educational data mining and learning analytics in programming: Literature review and case studies. In: Proceedings of the 2015 ITiCSE on Working Group Reports, Annual Conference on Innovation and Technology in Computer Science Education, pp. 41–63. ACM (2015). https://tutcris.tut.fi/portal/en/publications/educational-data-mining-and-learning-analytics-in-programming-literature-review-and-case-studies(6cd8ff1c-133a-4cf9-8a6e-ef61ba37ae7a).html
22. Fatima, D., Fatima, S., Prasad, A.V.K.: A survey on research work in educational data mining. J. Comput. Eng. **17**(2), 43–49 (2015)
23. Papamitsiou, Z., Economides, A.: Learning analytics and educational data mining in practice: a systematic literature review of empirical evidence. Educ. Technol. Soc. **17**(4), 49–64 (2014)
24. Isha, S., Dinesh, K., Mudit, K.: A review of applications of data mining techniques for prediction of students' performance in higher education. J. Stat. Manage. Syst. **20**(4), 713–722 (2017). https://www.tandfonline.com/doi/abs/10.1080/09720510.2017.1395191
25. Raheela, A., Agathe, M., Syed Abbas, A., Najmi, G.H.: Analyzing undergraduate students' performance using educational data mining. Comput. Educ. **113**, 177–194 (2017)
26. Anwar, A.Y., Addin, O.: Automatic classification of questions into Bloom's cognitive levels using support vector machines. In: The International Arab Conference on Information Technology. Naif Arab University for Security Science (NAUSS), Riyadh, Saudi Arabia (2013)
27. Anwar, A.Y., Addin, O., Mohammad, S.E.: Rocchio algorithm-based particle initialization mechanism for effective PSO classification of high dimensional data. Swarm Evol. Comput. **34**, 18–32 (2017). https://www.sciencedirect.com/journal/swarm-and-evolutionary-computation
28. Addin, O., Anwar, A., Y.: Classifications of exam questions using linguistically-motivated features: a case study based on Bloom's taxonomy. In: The Third International Arab Conference on Quality Assurance in Higher Education (IACQA 2016), pp. 889–896. Khartoum Sudan (2016)
29. Hamalainen, W., Vinni, M.: Comparison of machine learning methods for intelligent tutoring systems. In: ITS 2006 Proceedings of the 8th International Conference on Intelligent Tutoring Systems, Jhongli, Taiwan, pp. 525–534 (2006)
30. Mohamad, S.K., Tasir, Z.: Educational data mining: a review. In: The 9th International Conference on Cognitive Science, pp. 320–324. Procedia - Social and Behavioral Sciences, Kuching, Sarawak, Malaysia (2013)
31. Ronald, D.: The Importance of Having Data-sets. In: Proceedings of the IATUL Conferences, Paper 16 (2006)
32. Anwar, A.Y., Zakaria, T., Addin, O.: Bloom's Taxonomy–based classification for item bank questions using support vector machines. In: Modern Advances in Intelligent Systems and Tools, vol. 431, pp. 135–140 (2012). https://link.springer.com/book/10.1007/978-3-642-30732-4
33. Anwar, A.Y., Addin, O.: Automatic classification of questions into Bloom's cognitive levels using support vector machines. In: The International Arab Conference on Information Technology, pp. 335–342. Naif Arab University for Security Science (NAUSS), Riyadh, Saudi Arabia (2011). https://scholar.google.com/scholar?oi=bibs&cluster=11863385617269352176&btnI=1&hl=en

34. Anwar, A.Y., Addin, O., Ahmed A.A.: Educational data mining: a case study of teacher's classroom questions. In: 13th International Conference on Intelligent Systems Design and Applications (ISDA), pp. 34–41. UPM, Selangor (2013). http://ieeexplore.ieee.org/abstract/document/6920714/

35. Koller, D., Sahami, M.: Hierarchically classifying documents using very few words. In: International Conference on Machine Learning (ICML 1997), Nashville, Tennessee, pp. 170–178 (1997)

36. Weigend, A.S., Wiener, E.D., Pedersen, J.O.: Exploiting hierarchy in text categorization. Inf. Retrieval $\mathbf{1}$(3), 193–216 (1999)

37. Steven, B., Ewan, K., Edward, L.: Natural Language Processing with Python, 1st edn. O'Reilly Media, USA (2009)

38. Tsoumakas, G., Katakis, I., Vlahavas, I.: Mining multi-label data. In: Data Mining and Knowledge Discovery Handbook, pp. 667–685. Springer, Heidelberg (2010)

39. Jesse, R., Peter, R., Bernhard, P., Geoff, H.: MEKA: a multi-label/multi-target extension to Weka. J. Mach. Learn. Res. $\mathbf{17}$(21), 1–5 (2016)

40. Witten, I.H., Frank, E.: Data Mining: Practical Machine Learning Tools and Techniques, 2nd edn. Morgan Kaufmann, Elsevier, Amsterdam (2005)

41. Read, J., Pfahringer, B., Holmes, G.: Multi-label classification using ensembles of pruned sets. In: 8th IEEE International Conference on Data Mining, Pisa, Italy, pp. 995–1000. IEEE Computer Society (2008)

42. Sajnani, H., Javanmardi, S., McDonald, D.W., Lopes, C.V.: Multi-label classification of short text: a study on wikipedia barnstars. In: Analyzing Microtext: the Proceeding of the 2011 AAAI Workshop (2011)

# Software/Frameworks for Machine Learning and Data Mining Applications (Software)

# Bug Reports Evolution in Open Source Systems

Wajdi Aljedaani[1(✉)] and Yasir Javed[2,3(✉)]

[1] Al-Kharj College of Technology, Al-Kharj, Saudi Arabia
waljedaani@tvtc.gov.sa
[2] Network Security Research Group,
Faculty of Computer Science and Information Technology,
Universiti Malaysia Sarawak, Kota Samarahan, Malaysia
[3] Prince Sultan University, Riyadh, Saudi Arabia
yjaved@psu.edu.sa

**Abstract.** Open Source Software communities usually utilize open bug reporting system to enable users to report and fix bugs. In addition, the lifetime of most open source system stays for long periods of time. In this work, we comprehensively examine the evolution of bug reports in four different open source systems from various languages. The selected project are analyzed since 2004 in order to find how many bugs are reported compared to their resolution. We report our results and some recommendations to the open source community.

**Keywords:** Bug repository · Bug report · Open-source system bugs

## 1 Introduction

Large software systems are becoming essential to people daily lives. A big portion of these software systems is attributed to their maintenance. Several studies demonstrated that more than 90% of the software development cost is dedicated to maintenance and evolution tasks [1]. With a huge number of software systems, understanding and fixing bugs come to be a puzzling task. As bug reports (BRs) can facilities useful information to fix bugs, researchers have focused on how to leverage them to ease the task of fixing these bugs. The bug reports quality is very essential in easing the fix for that bug. Hooimeijer and Weimer [2] formed a descriptive model for bug reports lifetime. Bettenburg et al. [3] examined what makes a good bug report. Several researchers also explored different aspects of bug reports such as duplicate detection [4], fixer recommendation [5], feature prediction [6], and bug localization [7]. Bhattacharya et al. [8] investigated the fixing time of bug in Android apps. They studied both quality of bug reports and the bug-fixing process. A large amount of the software maintenance is attributed to bug fixing. Normally, bugs are reported, fixed, verified and closed. Nevertheless, in some scenarios, bugs

© Springer International Publishing AG, part of Springer Nature 2018
M. Alenezi and B. Qureshi (Eds.): *5th International Symposium on Data Mining Applications*, pp. 63–73, 2018.
https://doi.org/10.1007/978-3-319-78753-4_6

have to be re-opened, which will increase maintenance costs, degrade the overall user-perceived quality of the software and lead to unnecessary rework.

Bug fixing is a very important essential activity in software development and maintenance. Bugs are reported, recorded, and managed in bug tracking systems such as Bugzilla. A bug report contains many fields, which provide insights into the bug and help in fixing it. Open source software (OSS) development is a collaborative large team activity contributed by many globally distributed developer community. Many open source software teams depend on freeware tools to plan, code, test, track-report-fix bugs and market product(s). Open source projects embrace open bug repositories such as Bugzilla to support its development and maintenance in managing and maintaining bugs. Geographically-distributed users and developers can report software bugs. This will help in addressing these bug reports. In this way, open source software is iteratively developed and the quality of the produced software can be improved [4]. A bug triage is the person who will decide to whom a bug reported is assigned.

There is a number of reasons for bugs in open source systems but their resolution requires a proper attention from developers and contributors of the project. In order to understand how bugs are reported over time and how bugs are resolved the authors tend to perform this research. This research will also highlight the time taken to resolve the bugs that might vary from days to months but in some case more than a certain number of years. The authors will also propose a model to avoid a situation where a low priority bug always sits unresolved. Sometimes bugs are often misreported and thus require no attention, this study will also look into how many of this kind of case exists.

## 2    Literature Review

A number of researchers have studies the qualitative analysis of bug repositories on a survey, and open software system [9–12]. Bettenburg et al. [13] examined bugs performance in two software systems Firefox and chrome to study the collaboration among project team members to identify performance bugs. Their investigation centered done on Firefox for bugs that on Firefox and Firefox core component, and the entire bug reports that were submitted to chrome system.

Banerjee et al. [9] studied two bug repositories of Mozilla and Eclipse in order to compare the similarities and differences between each of them. Their goal was to evaluate user behavior, the structure of the repository and the type of bugs. They also looked into the duplicating groups as well as the frequency of the reported bugs. Our study is similar to their in the frequency of reported bug. They concentrated only on two open source software Mozilla and Eclipse, however, we studied four open source software Ant, K3B, Kate, and OpenSSH. We analyzed the evolution of bug repository for these projects over the time reporting bug per year. Bugs evolve over time and in order to make them maintainable and making sure that they stay with passage over time require the resolution of bugs as mentioned by [14]. The authors in [15] worked in the classification of bugs in order to find out the impact of bugs but were limited to manual checking and only Apache projects. The authors in [16] have worked on the system of bounty for awarding problem solvers to award them with some gifts or monetary values. It also highlights the need for funding for open source projects to sustain in the market and thus this study will highlight the issues in terms of days to resolve the project. It will also highlight that commitment in projects make the project running else open source projects community consist of a number of project but are not being used due to lack of support.

# 3   Bug Tracking System

When a developer or a user encountered a bug while using the software or wants to request an enhancement, he or she usually open a bug report in the open bug repository. Many open bug repositories (e.g., Bugzilla, Jira, Gnats, and Trac) have been adopted in open source projects. We only explore projects that use Bugzilla as their bug tracking system. This Section introduces background material necessary for this work.

In Bugzilla, bugs are kept in the form of bug reports which consist of predefined fields, text description, and attachments. Bugzilla keeps a record of several information about bug reports in its database. The fields are usually found are:

- Bug ID: Unique identifier for bug reports.
- Description: Textual description of the bug.
- Opened: Date of the report.
- Status: Status of the report. (new, assigned, reopened, needinfo, verified, closed, resolved and unconfirmed).
- Resolution: Action to be performed on the bug (obsolete, invalid, incomplete, notgnome, notabug, wontfix, and fixed).
- Assigned: Name and/or e-mail address of the developer in charge of fixing this bug.
- Priority: Urgency of the error. (immediate, urgent, high, normal and low).
- Severity: How this error affects the use and development of the software. (blocker, critical, major, normal, minor, trivial and enhancement).
- Reporter: Name and e-mail address of the bug reporter.
- Product: Software that contains the bug.
- Version: Version number of the product.
- Component: a Minor component of the product.
- Platform: Operating system or architecture where the error appeared.

Figure 1 shows the bug life cycle adapted form Bugzilla, whenever a new bug is reported, it report is forwarded to the developer for deciding whether the bugs need further details or whether it will be fixed. Once the bug is accepted the Bugs is fixed and the report goes to the reporter. If the solution is accepted then the bug is considered to be resolved else the bug is reopened and goes back to developer.

## 4    Data Collection

For a selected software program, we extracted the vital data from the bug repository. From that point, we identify the bug reports that were relevant to our software system and perform an extraction of metric for the entire bug reports in all of the four projects. This research concentrated on the bug reports metrics to identify the similarities and differences between each project in our dataset. We gathered the data of all bug reports in all projects that were submitted to bug repository "Bugzilla" in the period from 2004 to 2017 and retrieved almost 13,047 bugs. In this study, we collect all types of bugs, for example, closed, opened, wont fix, etc. First, we check all the available bug repositories for all examined software system in the Bugzilla bug tracker website. For each software system, we extract the essential data from the bug repository (i.e., bugID, description, status, resolution, priority, and reporter). As shown in Fig. 2 we have used Bugzilla from three major sources (1) Eclipse (2) KDE and (3) OpenSSH. This research then extracted all the bugs to conduct the study.

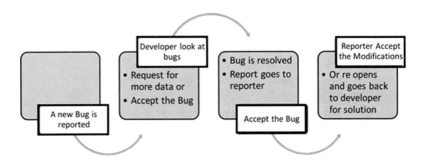

**Fig. 1.** Bugs life cycle as reported by Bugzilla

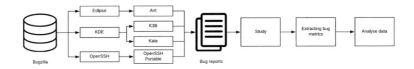

**Fig. 2.** Process of how this research is conducted

## 5    Research Methodology

This research focused on open source projects from various languages that are Ant [17], K3B [18], Kate [19] and OpenSSH [20]. We have looked over these projects from their first bug reporting year. These projects are evaluated in term of Reported bugs, Number of unsolved bugs, Number of latest commits, Number of Duplicated Bugs and Average time to fix the bugs.

Each project bug has been extracted over the years till date using the crawlers built by us. It is seen that OpenSSH started with the highest number of bugs but has decreased a lot to 213 while for ANT has reduced the number of bugs to half. The most number of bugs were for ANT was 2402 in 2008. Table 1 shows the detail about the system that was selected.

**Table 1.** Project details along with their bugs.

| System | Platform | Language | #Bug reports |
|--------|----------|----------|--------------|
| Apache | Ant | Java | 3,608 |
| KDE | K3B | C++ | 1,838 |
| KDE | Kate | C++ | 4,902 |
| OpenSSH | OpenSSH portable | C | 2,699 |
| **Total** | **13,047** | | |

Table 2 shows the number of bugs collected for each project from the year 2004 to 2017 in terms of a total number of bugs for each year and number of fixed bugs for each year.

**Table 2.** The bugs reported and bugs solved since 2004 for all the projects.

| Year | # Bug reports | | | | # Fixed bugs | | | |
|------|-----|-----|------|---------|-----|-----|------|---------|
|      | Ant | K3B | Kate | OpenSSH | Ant | K3B | Kate | OpenSSH |
| 2004 | 102 | 7 | 366 | 764 | 94 | 0 | 209 | 379 |
| 2005 | 49 | 272 | 171 | 32 | 44 | 115 | 89 | 16 |
| 2006 | 21 | 410 | 99 | 263 | 17 | 265 | 40 | 129 |
| 2007 | 10 | 105 | 149 | 18 | 3 | 55 | 77 | 4 |
| 2008 | 2402 | 72 | 163 | 216 | 1044 | 25 | 77 | 137 |
| 2009 | 252 | 88 | 310 | 103 | 106 | 56 | 111 | 43 |
| 2010 | 175 | 434 | 696 | 157 | 87 | 43 | 299 | 94 |
| 2011 | 96 | 59 | 534 | 164 | 30 | 25 | 217 | 80 |
| 2012 | 112 | 29 | 570 | 22 | 50 | 4 | 214 | 1 |
| 2013 | 97 | 13 | 400 | 48 | 42 | 5 | 151 | 21 |
| 2014 | 133 | 224 | 293 | 89 | 72 | 11 | 123 | 35 |
| 2015 | 60 | 22 | 404 | 279 | 14 | 0 | 103 | 118 |
| 2016 | 45 | 51 | 450 | 331 | 21 | 15 | 102 | 147 |
| 2017 | 54 | 52 | 297 | 213 | 13 | 21 | 40 | 2 |

## 6   Analysis and Discussion

As shown in Fig. 3, the total numbers of resolved bugs are mostly done for Ant except for the year 2008, where actually the numbers of reported bugs were too much. A number of solved bugs reveal that resolution and dedication of committers with the project and overall success of the project.

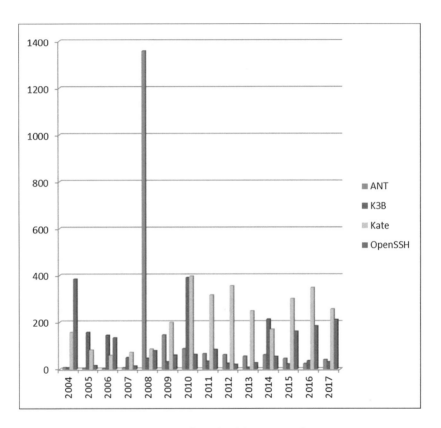

**Fig. 3.** Number of resolved bugs over the year

Figure 3 shows only the bugs resolved, but most of these bugs were resolved in short time while other were closed as they were wrongly reported. As shown in Table 3, most numbers of bugs are not open except ANT that has 10 opened in 2009. This shows that committers tend to solve most bugs and fix them. [14] also shows that bugs are sometimes misreported and requires a proper mechanism for developers to close this kind of issues. It also shows that some bugs need not be fixed and these fixings requires either some additional effort and will be fixed in upcoming releases. Except Ant and Kate, OpenSSH and K3B have less won't fixed bugs referring to their popularity and their commitment to solve problem. Moreover, the community in OpenSSH is big it also refers to the quick solution of bugs as more dedication is offered by a most number of peoples.

**Table 3.** Number of opened as well as won't fixed bugs.

| Year | # Opened bugs | | | | # Won't fixed bugs | | | |
|------|-----|-----|------|---------|-----|-----|------|---------|
|      | Ant | K3B | Kate | OpenSSH | Ant | K3B | Kate | OpenSSH |
| 2004 | 0   | 0   | 0    | 0       | 1   | 0   | 17   | 65      |
| 2005 | 0   | 0   | 0    | 1       | 1   | 5   | 11   | 1       |
| 2006 | 0   | 0   | 0    | 1       | 0   | 11  | 5    | 23      |
| 2007 | 0   | 0   | 0    | 0       | 1   | 1   | 11   | 0       |
| 2008 | 9   | 0   | 0    | 0       | 220 | 1   | 16   | 32      |
| 2009 | 10  | 0   | 0    | 1       | 11  | 0   | 7    | 15      |
| 2010 | 1   | 0   | 0    | 1       | 6   | 0   | 20   | 18      |
| 2011 | 1   | 0   | 0    | 2       | 6   | 0   | 43   | 27      |
| 2012 | 1   | 0   | 0    | 0       | 9   | 0   | 101  | 1       |
| 2013 | 1   | 0   | 0    | 2       | 21  | 0   | 25   | 1       |
| 2014 | 3   | 2   | 1    | 1       | 5   | 4   | 18   | 0       |
| 2015 | 1   | 0   | 0    | 7       | 9   | 0   | 113  | 19      |
| 2016 | 1   | 2   | 6    | 2       | 0   | 1   | 45   | 40      |
| 2017 | 2   | 1   | 3    | 2       | 4   | 0   | 7    | 1       |

A number of duplicated bugs refers to the point where bugs might be reported differently by two or more people. The opened bugs are referred to committers and they mostly decide whether the bug is duplicate or not. A number of duplicated bugs also refers to how many people faced the same issue and may require more attention than other bugs as shown in Fig. 4.

It is seen that ANT has the only spike of high duplicated bugs with Kate have the same issue at number of years but is tend to reduce in current years that shows the awareness among developers regarding the solution of the problems as shown in Eq. 1. Equation 1 allows to find a predictor based on trend about number duplicated bugs in upcoming years.

$$y = -25.91 \ln x + 60.98 \tag{1}$$

A total number of commits represent the dedication of developers and contributors dedication toward the success of the project as shown in Fig. 5, shows the number of bug reporters has decreased over the current years that also make sure about the project stability and less number of bugs. It may also depict that project are being fixed consistently and thus require less number of bugs.

The number of days per solution to bug varies from 0 to all four projects to above 5K days that may be due to non-priority of the bug or the bugs require fixing some menu issues and that is solved after the solution of each bug. The average of each bug solving is highest in ANT that is around 1125 days while lowest in K3B that means each bug is solved in a less than year calendar while Kate is also solved in an year as shown in Table 4.

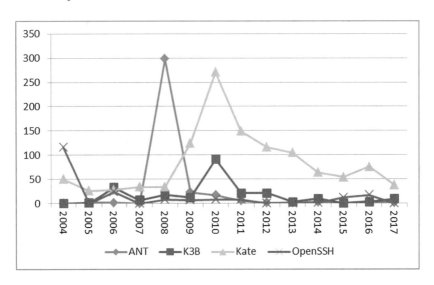

**Fig. 4.** The number of duplicated bugs over the years for all four projects

While looking at above results it is proposed that a complete reporting should be made that can help with following:

**Table 4.** Summary of fixing time of the projects (in #days).

| Platform | #Minimum | #Average | #Maximum |
|----------|----------|----------|----------|
| Ant | 0 | 1125.061 | 4721 |
| K3B | 0 | 274.54 | 2854 |
| Kate | 0 | 370.22 | 4857 |
| OpenSSH | 0 | 572.94 | 5470 |

- Auto classification of bug when reported. This will require two approaches to be followed (1) the proposed system can ask reported to bug to select the type of bug (2) second and most important approach is text mining and auto classification of bug according the reported issue. This means that whole reported text will be matched against bug classification corpus already built and will be assigned a weight for each category. The category with highest weight will be selected.
- Assigning it to relevant developer automatically that means checking about which developer can fix similar kind of bug, which developer have less load, which developer needs less time to fix the bug thus will be dependent on number of factors and all will be taken into account.
- Assigning the time-line to fix the bug depending on its category like a programmatic ally bug error will have a small fixing time compared to cosmetic changes.

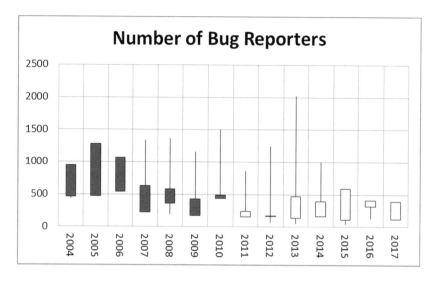

**Fig. 5.** The number of bug reporters each year combined for a project

- Propose developers already known solution for the reported kind of bug in order to expedite the process. This will help the developers in doing the quick fix and will allow the developer to accept or reject the solution in order to help other developers whether the suggested solution is better or not.
- Once the bug is solved, a message about fixing the bug will be sent to reported in order to verify and close the ticket. This is will an accountability and easy quality assurance system.

## 7    Conclusion

This paper looks into the number of systems that are Open Source and popularly used by industry and academia. The authors selected top most used system from different categories all built in different languages. These systems are seen for their number of bug reported by people, the time it takes to solve the bugs and how many bugs are solved over the calendar year. The bugs reporting show that there is a number of people that are using the system and efficiently contributing to testing an open source system that shows the reliability of the system. If the bugs are not solved in time the system fails and lacks interests but continuous development and contribution assures that selected system are still in use since 2004. The authors suggest making a system in order to highlight the bugs to committers and contributors. The system will help in classification of bugs as well as make sure that none of the bugs is active and unresolved in a calendar year. The authors introduced an aging concept for bugs so that the lowest priority bugs make a high priority to solve the problem of overlooking. In future, authors plan to look into the bugs and classify them according to their type thus making the proposed system reality. The system will automatically look for solution using crawlers to find the right kind of solution.

# References

1. Shihab, E., Ihara, A., Kamei, Y., Ibrahim, W.M., Ohira, M., Adams, B., Hassan, A.E., Matsumoto, K.: Studying re-opened bugs in open source software. Empir. Softw. Eng. **18**(5), 1005–1042 (2013)
2. Hooimeijer, P., Weimer, W.: Modeling bug report quality. In: Proceedings of the Twenty-second IEEE/ACM International Conference on Automated Software Engineering, pp. 34–43. ACM (2007)
3. Zimmermann, T., Premraj, R., Bettenburg, N., Just, S., Schroter, A., Weiss, C.: What makes a good bug report? IEEE Trans. Softw. Eng. **36**(5), 618–643 (2010)
4. Alipour, A., Hindle, A., Stroulia, E.: A contextual approach towards more accurate duplicate bug report detection. In: Proceedings of the 10th Working Conference on Mining Software Repositories, pp. 183–192. IEEE Press (2013)
5. Xie, X., Zhang, W., Yang, Y., Wang, Q.: Dretom: developer recommendation based on topic models for bug resolution. In: Proceedings of the 8th International Conference on Predictive Models in Software Engineering, pp. 19–28. ACM (2012)
6. Tian, Y., Lo, D., Sun, C.: Information retrieval based nearest neighbor classification for fine-grained bug severity prediction. In: 2012 19th Working Conference on Reverse Engineering (WCRE), pp. 215–224. IEEE (2012)
7. Saha, R.K., Lease, M., Khurshid, S., Perry, D.E.: Improving bug localization using structured information retrieval. In: 2013 IEEE/ACM 28th International Conference on Automated Software Engineering (ASE), pp. 345–355. IEEE (2013)
8. Bhattacharya, P., Ulanova, L., Neamtiu, I., Koduru, S.C.: An empirical analysis of bug reports and bug fixing in open source android apps. In: 2013 17th European Conference on Software Maintenance and Reengineering (CSMR), pp. 133–143. IEEE (2013)
9. Banerjee, S., Helmick, J., Syed, Z., Cukic, B.: Eclipse vs. Mozilla: a comparison of two large-scale open source problem report repositories. In: 2015 IEEE 16th International Symposium on High Assurance Systems Engineering (HASE), pp. 263–270. IEEE (2015)
10. Zibran, M.F., Eishita, F.Z., Roy, C.K.: Useful, but usable? Factors affecting the usability of APIs. In: 2011 18th Working Conference on Reverse Engineering (WCRE), pp. 151–155. IEEE (2011)
11. Ko, A.J., Chilana, P.K.: Design, discussion, and dissent in open bug reports. In: Proceedings of the 2011 iConference, pp. 106–113. ACM (2011)
12. Bettenburg, N., Just, S., Schröter, A., Weiß, C., Premraj, R., Zimmermann, T.: Quality of bug reports in eclipse. In: Proceedings of the 2007 OOPSLA Workshop on Eclipse Technology eXchange, pp. 21–25. ACM (2007)
13. Bettenburg, N., Just, S., Schröter, A., Weiss, C., Premraj, R., Zimmermann, T.: What makes a good bug report? In: Proceedings of the 16th ACM SIGSOFT International Symposium on Foundations of Software Engineering, pp. 308–318. ACM (2008)
14. Javed, Y., Alenezi, M.: Defectiveness evolution in open source software systems. Procedia Comput. Sci. **82**, 107–114 (2016)
15. Wang, J., Keil, M., Oh, L., Shen, Y.: Impacts of organizational commitment, interpersonal closeness, and confucian ethics on willingness to report bad news in software projects. J. Syst. Softw. **125**, 220–233 (2017)

16. Karim, M.R., Ihara, A., Yang, X., Iida, H., Matsumoto, K.: Understanding key features of high-impact bug reports. In: 2017 8th International Workshop on Empirical Software Engineering in Practice (IWESEP), pp. 53–58. IEEE (2017)
17. Apache ant - welcome. http://ant.apache.org/. Accessed 11 Feb 2018
18. Github - kde/k3b: K3b is a full-featured cd/dvd/blu-ray burning and ripping application. https://github.com/KDE/k3b. Accessed 11 Feb 2018
19. Github - kde/kate: An advanced editor component which is used in numerous kde applications requiring a text editing component. https://github.com/KDE/kate. Accessed 11 Feb 2018
20. OpenSSH. https://www.openssh.com/. Accessed 11 Feb 2018

# Analysis of Call Detail Records for Understanding Users Behavior and Anomaly Detection Using Neo4j

Emsaieb Geepalla[1]([⊠]), Nasser Abuhamoud[1]([⊠]),
and Abdulla Abouda[2]([⊠])

[1] School of Communication and Computer Engineering,
Sebha University, Sebha, Libya
{Ems.geepalla,Mans.abuhamoudl}@sebhau.edu.ly
[2] Research and Development Office, Almadar Aljadid Co., Tripoli, Libya
A.abouda@almadar.ly

**Abstract.** Call Detail Records (CDR) is a valuable source of information; it opens new opportunities for mobile operator industries and maximize their revenues as well as it helps the community to raise its standard of living in many different ways. Nevertheless, we need to analyze CDR in order to extract its big value and detect abnormal costumers behaviors to help companies to develop their future plans. However the analysis of CDRs is a very complex process this because it has a huge volume of data. Therefore, In this paper we propose an approach that makes use of Neo4j for automatic analysis of CDRs. To achieve this we transformed the CDR data into neo4j and then we used cypher query language for performing an automatic analysis. A real case study was used to evaluate the proposed approach.

**Keywords:** Call Detail Records · Big data · Neo4j · Graph database
Abnormal user behavior

## 1 Introduction

Today the popularity and wide diffusion of cellular phones, a huge quantity of mobile devices are moving everyday with their human companions, leaving tracks of their movements and their everyday habits. Mobile phones are becoming pervasive in both developed and developing countries and they can be a precious source of data and information, with a significant impact on research in behavioral science [1].

A Call Data Record (CDR) is a data structure storing relevant information about a given telephonic activity involving an user of a telephonic network. A CDR usually contains spatial and temporal data and it can carry other additional useful information. Population census have been widely used in the past for keeping track of the demography and geographical movements of the population [1–3]. Nowadays, due to short term and everyday mobility, more flexible methods such as various registers and indirect databases are employed CDRs represent an optimal candidate in this sense. One of their main advantage is that they offer a statistically accurate representation of

© Springer International Publishing AG, part of Springer Nature 2018
M. Alenezi and B. Qureshi (Eds.): *5th International Symposium on Data Mining Applications*, pp. 74–83, 2018.
https://doi.org/10.1007/978-3-319-78753-4_7

the distribution of people in an area and they can be used to track large and hetero-geneous groups of people [4–6]. Since CDRs evolve accordingly to the changes of user's behavior, the information they carry "automatically" updates over time. Telecom operators continuously gather a huge quantity of CDRs, from which it is possible to extract additional information with low additional costs and generate valuable datasets. Analyses of CDR data can be successfully employed in many different fields, like anomaly detection, monitoring the network, Simbox detection, adaptation of supplied services (e.g., customers' billing, network planning) and understanding of the eco-nomic level of a certain area [1, 7].

The focus of this paper is to analyze CDRs in order to understand users' behaviors and detect abnormal user behavior. However the analysis of such data is considered a vital task this is due to the large amount of data generated every minute. Therefore, this paper make use of graph-based technology for performing an automatic analysis on CDRs. More precisely, in this paper we transform CDRs data into Neo4j and perform the automatic analysis with the help of cypher query language. Neo4j has been used because it provides cost effective and scalable solutions capable to process Big Data with the required per-formance measures and unlike many other big data technology neo4j has the capability of visualization [8]. The proposed approach has been evaluated with the help of real case study provided by a tier-1cellular operator in Libya (Almadar Aljadid Co.,) [9].

The remainder of this paper is organized as follows. Section 2 provides a review of CDRs, Neo4j and cypher query language. Section 3 presents a description of the problem. In Sect. 4 we briefly describe the running example, Sect. 5 illustrates how CDR data imported into neo4j. Section 6 presents several abnormal scenarios and describes the analysis of the CDR data using cypher. Finally the paper ends with a conclusion in Sect. 7.

## 2   Preliminary

This section provides a brief introduction to CDRs, Neo4j and Cypher query language.

### 2.1   Call Detail Records

A call detail record (CDR) is a data record produced by a telephone exchange or other telecommunications equipment that documents the details of a telephone call or other telecommunications transaction (e.g., text messages) and any other official communi-cations transmission. that passes through that facility or device [4, 10]. The record contains various attributes of the call, such as call duration, start time, completion status, calling number, and called number [5]. The call detail record simply shows that the calls or messages took place, and measures basic call properties.

## 2.2  Neo4j

Neo4j is the implementation chosen to represent graph databases. It is open source for all noncommercial uses. It has been in production for over five years. It is quickly becoming one of the foremost graph database systems. According to the Neo4j website, Neo4j is "an embedded, disk-based, fully transactional Java persistence engine that stores data structured in graphs rather than in tables" [11, 12]. The developers claim it is exceptionally scalable (several billion nodes on a single machine), has an API that is easy to use, and supports efficient traversals. Neo4j is built using Apache's Lucene 3 for indexing and search. Lucene is a text search engine, written in Java, geared toward high performance [8].

## 2.3  Cypher Query

Cypher is an expressive (yet compact) graph database query language. Cypher is designed to be easily read and understood by developers, database professionals, and business stakeholders. Its ease of use derives from the fact that it is in accord with the way we intuitively describe graphs using diagrams [13]. Cypher enables a user (or an application acting on behalf of a user) to ask the database to find data that matches a specific pattern [11].

## 3  Description of the Problem

Call monitoring and recording applications used by telecommunication companies generate extremely large amount of call detail records (CDRs) in real-time, and companies constantly need to leverage from this data to boost productivity. The volume of the calls and data captured by the call monitoring applications is very huge that it impossible to manually analyze and conclude the behavior of customers or network.

When "abnormal customers' behaviors" occur infrequently but often enough, they can cause economic loss and negatively affect productivity and efficiency of a thriving business. For example, when a user is making a high number of calls and does not receive any calls this might give intuition that the user's account is a fraudulent Simbox account. Fraudulent Simbox hijacks international voice calls and transfer them over the Internet to a cellular device, which injects them back into the cellular network. As a result, the calls become local at the destination network [14], and the cellular operators of the intermediate and destination networks do not receive payments for the call routing and termination Thus, the analysis of CDRs in order to understand users' behavior and detect abnormal behavior is considered as a vital task. Therefore, in this paper we propose a method that transforms the CDRs data into Neo4j, thus allowing for powerful automatic analysis, in order to understand the behaviors of customers and detect abnormal customers' behaviors.

## 4    Running Example

In this section we briefly describe the running example that has been used to evaluate our approach. The case study that we used is a real case study provided by a tier-1cellular operator in Libya (Almadar Aljadid Co.,) and the data is collected between 01/01/2016 to 01/04/2016. However for the analysis conducted in this study we have only used the columns (CDR features) that we are interested in. These features are: calling number, called number, location of calling number, location of called number, call duration, and exchange.

## 5    Transformation from CDRs into Neo4j

Our work motivated by the need to bridge the gap between raw date and graph database (Neo4j) in order to allow for a powerful analysis to be carried out using cypher query language. Figure 1 depicts an overview of our approach to transform CDRs data into Neo4j. More precisely, the figure shows that we firstly need to design a graph model to represent a phone call. Secondly we make use of cypher query language to import the CDRs data into Neo4j. The produced graph database can be visualized and analyzed via cypher query language. Next we shall describe the transformation phases in more details.

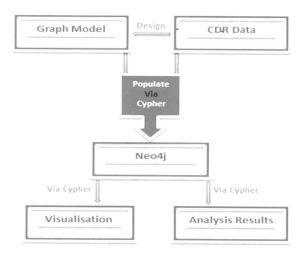

**Fig. 1.**  Architecture of our approach

The process of transforming the CDRs data into Neo4j consists of two steps which are described as follows.

**Step 1:** Defining the graph model that represent a phone call.
The following figure (Fig. 2) depicts the graph model that represents a CDR record. It illustrates how the CDR features are represent in the graph.

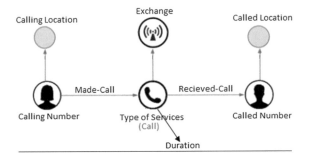

**Fig. 2.** Graph model

**Step 2:** Using cypher query language to populate CDR data into Neo4j.

To transform the CDR data into neo4j we have created several cypher queries. Figure 3 presents samples of the cypher queries that have created to generate the appropriate nodes. Using the Cypher queries depicted in the following figure, we'll create a node for each calling number, called number, location of calling number, location of called number, call duration, exchange and type of services.

```
CREATE (CallinNode:Make_Call { name:'Calling' })
SET CallinNode = ROW,
CallinNode.callingPartyNumber = (ROW.callingPartyNumber)
CREATE (CalledNode:Reciever_Call { name:'Called' })
SET CalledNode = ROW,
CalledNode.calledPartyNumber= (ROW.calledPartyNumber)
CREATE (Exchange:Exchange_Identity{name:'Exchange'})
SET Exchange = ROW,
Exchange.Exchange_Identity=(ROW.exchangeIdentity)
CREATE (f:Location { name:'Location' })
SET f = ROW,
f.Location = (ROW.lastCallingLocationInformation)
CREATE (g:Located { name:'Located_from' })
SET g = ROW,
g.Located = (ROW.lastCalledLocationInformation)
CREATE (s:duration{ name:'DURATION'})
SET s = ROW,
s.duration = (ROW.chargeableDuration)
CREATE (CALL:Call{NAME:'CALL'})
```

**Fig. 3.** Cypher queries for nodes creation

The created nodes should be linked by relationships. Therefore we have generated several cypher queries to achieve this. For example the cypher queries presented in Fig. 4 will generate relationships between calling number, calls and called number, as well as between calls, exchange and duration.

```
CREATE (CallinNode)-[:Made_Call]->(CALL)-[:received_call]->(CalledNode),
(f)-[:FRM_BTS]->(CallinNode), (g)-[:TO_BTS]->(CalledNode).
CREATE (CallinNode)-[:Made_Call]->(CALL)-[:received_call]->(CalledNode),
(Exchange)-[:FROM]->(CALL), (s)-[:DURATION]->(CALL).
```

**Fig. 4.** Cypher queries for relationships creation

# 6 Analysis of the CDRs Data Using Neo4j and Cypher

In the previous section, we have shown the transformation of the CDRs in raw format into graph database. However, this work would be incomplete without demonstrating how we can analyze the produced graph database to help the mobile operator company to understand their customers behaviors and detect anomaly. For this purpose we have generated several cypher queries. Before starting the automatic verification to detect abnormal user behavior, we must establish what is meant by abnormal in CDRs data. Next we provide several scenarios that could be considered as abnormal behavior of users.

## 6.1 Abnormal User Behavior

User behavior is considered as abnormal if the user behavior is significantly different from that normally observed. In our approach this means we look at the CDR data to see whether the user has done something that is unusual or unexpected. The following are scenarios that could be considered as abnormal user behavior.

- Abnormal based on duration
  If a user is making calls which is longer than n, where n is defined by the mobile operator company. Detecting such scenarios might help mobile operator companies for Simbox detection.
- Abnormal based on movement
  In this scenario we will check whether we can find any user who is making all his/her calls from one location.
- Abnormal based on type of services
  This means we will check whether there is any users who is only using call services.
- Abnormal based on high number of calls
  In this scenario will find out the list of the users who are making a high numbers of calls.

Detecting all the above scenarios might help mobile operator companies during the process of Simbox detection. This is because most of Simbox are not moving, making, high number of calls, and making calls with long durations.

## 6.2    Experimental Result

In this section we describe how CDR data could be analyzed using Neo4j and Cypher. To do so we have formalized several query using cypher query language. The following statement is an example of the cypher queries that we have formalized (Fig. 5).

```
MATCH (a)-[r]-(b)
return distinct a.Receiver_Call as number_Received, count(r) as Numbers_of_calls
order by Numbers of calls desc
```

**Fig. 5.** Query for number of calls

The execution of the above query provides us a list of the users who have made the highest number of calls and did not receive any call. Figure 6 depicts a short list of those numbers. It shows that the user who has number A1 has made the highest number of calls with 2753 calls. Making such high number of calls while did not receive any call during the same period of time require more investigation by the company to make sure that such users are not Simbox.

**Fig. 6.** Number of calls for every user

Another example of the cypher queries that we have generated is illustrated in Fig. 7.

```
CREATE (a:Make_Call { name:'Calling' })

SET a = ROW, a.Made_Call = toint(ROW.Made_Call)

return distinct a.From Locationa as Tower, count(a.Made Call) as Call Out order by Call Out desc
```

**Fig. 7.** Query for number of locations that users' visited

The execution of this cypher query provides us with information about the number of locations that users visit when they made their calls, as depicted in Fig. 8. It illustrates that there is 0.002% of the whole users are making all their calls from one location. Therefore, the company has to ensure that this percentage of users is not Simbox by conducting more investigation.

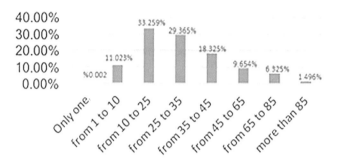

**Fig. 8.** Number of locations

Another example of cypher queries which we have formalized, in order to perform the analysis on the CDR data is the query depicted in Fig. 9.

```
MATCH (a)
where a.Duration_of_Call > "0:01:14"
return a.Made_Call as Made_Call, count(*) as counter order by counter desc
```

**Fig. 9.** Query for duration of calls

The execution of this query gives the percentage of users based on the duration of calls as depicted in Fig. 10. For example, it shows that 0.6% of the whole customers are making calls with long duration (30 min or longer). Those customers are likely to cause revenue loss to the company if they are Simbox. Hence, to avoid any loss might be caused by those users, the company needs to conduct more analysis to ensure that those users are not Simbox.

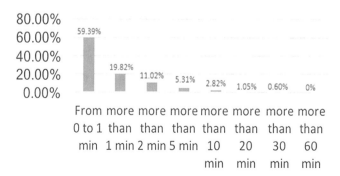

**Fig. 10.** Duration of calls

One more query that have been generated is illustrated in the following figure (Fig. 11).

```
CREATE (a:Make_Call { name:'Calling' })
SET a = ROW,   a.Made_Call = tofloat(ROW.MadeCall)
RETURN distinct  a.Exchange as Exchange,count(*) as counter order by counter desc order by Call  Out desc
```

**Fig. 11.**  Query for number of users based on exchanges

The execution of this cypher query provides us with the number of users for every exchange as shown in Fig. 12. The figure illustrates that the exchange MSC-BC1 is the busiest exchange with 29865643 users. Such information will help the company to plan for future development.

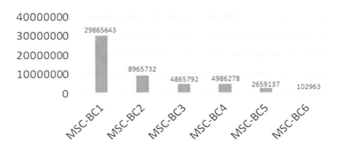

**Fig. 12.**  Number of users for every exchange

## 7   Conclusion

Using big data technology is always beneficial to model, check and verify large data. The main contribution of this paper is that it presents a graph-based method that makes use of big data technology to model and analyze CDRs data for understanding the user behavior and detect abnormal user behavior. In particular, it describes how can we transform the CDRs data from the raw format into graphs using Neo4j and Cypher, thus allowing for powerful analysis to take place using cypher queries. The feasibility of this approach was demonstrated through a real case study.

**Acknowledgments.**  This research was supported by the Research & Development (R&D) office at Almadar Aljadid Company. We thank our colleagues from R&D office at Almadar Aljadid Company who provided insight and expertise that greatly assisted the research.

# References

1. Hoteit, S., Chen, G., Viana, A., Fiore, M.: Filling the gaps: on the completion of sparse call detail records for mobility analysis. In: The Eleventh ACM Workshop on Challenged Networks, pp. 45–50 (2016)
2. Elagib, S.B., Hashim, A.A.-H., Olanrewaju, R.F.: CDR analysis using big data technology. In: International Conference on Computing Control Networking Electronics and Embedded Systems Engineering (ICCNEEE), pp. 467–471, September 2015
3. Kedma, G., Guri, M., Sela, T., Elovici, Y.: Analyzing users' web surfing patterns to trace terrorists and criminals. In: Intelligence and Security Informatics (ISI), pp. 143–145, June 2013
4. Fiadino, P., Ponce-Lopez, V., Antonio, J., Torrent-Moreno, M., D'Alconzo, A.: Call detail records for human mobility studies: taking stock of the situation in the "always connected era". In: Proceedings of the Workshop on Big Data Analytics and Machine Learning for Data Communication Networks, pp. 43–48. ACM (2017)
5. Pestre, G., Letouzé, E., Zagheni, E.: The ABCDE of big data: assessing biases in call-detail records for development estimates. In: Annual World Bank Conference on Development Economics (2016)
6. Zhao, Z., Shaw, S.-L., Xu, Y., Yin, L.: Understanding the bias of call detail records in human mobility research. Int. J. Geograph. Inf. Sci. **30**(9), 1738–1762 (2016)
7. Kim, H.K., Kim, T.E., Jo, C.M., Na, S.R., Jurn, J.S.: Abnormal behavior detection system considering error rate deviation of entire use behavior pattern during personalized connection period. U.S. Patent Application 15/006,498. Filed 26 January 2016
8. Vukotic, A., Watt, N., Abedrabbo, T., Fox, D., Partner, J.: Neo4j in action. Manning Publications Co., Greenwich (2014)
9. https://www.almadar.ly/
10. Kumar, M., Hanumanthappa, M., Suresh Kumar, T.V.: Crime investigation and criminal network analysis using archive call detail records. In: 2016 Eighth International Conference on Advanced Computing (ICoAC), pp. 46–50. IEEE (2017)
11. Miller, J.J.: Graph database applications and concepts with neo4j. In: Proceedings of the Southern Association for Information Systems Conference, Atlanta, GA, USA, vol. 2324, p. 36 (2013)
12. Van Bruggen, R.: Learning Neo4j. Packt Publishing Ltd, Birmingham (2014)
13. Holzschuher, F., Peinl, R.: Performance of graph query languages: comparison of cypher, gremlin and native access in Neo4j. In: Proceedings of the Joint EDBT/ICDT 2013 Workshops, pp. 195–204. ACM (2013)
14. Hung, S.-Y., Yen, D.C., Wang, H.-Y.: Applying data mining to telecom churn management. Exp. Syst. Appl. **31**(3), 515–524 (2006)

# Empirical Analysis of Static Code Metrics for Predicting Risk Scores in Android Applications

Mamdouh Alenezi[✉] and Iman Almomani

College of Computer and Information Sciences, Prince Sultan University,
Riyadh, Saudi Arabia
{malenezi,imomani}@psu.edu.sa

**Abstract.** Recently, with the purpose of helping developers reduce the needed effort to build highly secure software, researchers have proposed a number of vulnerable source code prediction models that are built on different kinds of features. Identifying security vulnerabilities along with differentiating non-vulnerable from a vulnerable code is not an easy task. Commonly, security vulnerabilities remain dormant until they are exploited. Software metrics have been widely used to predict and indicate several quality characteristics about software, but the question at hand is whether they can recognize vulnerable code from non-vulnerable ones. In this work, we conduct a study on static code metrics, their interdependency, and their relationship with security vulnerabilities in Android applications. The aim of the study is to understand: (i) the correlation between static software metrics; (ii) the ability of these metrics to predict security vulnerabilities, and (iii) which are the most informative and discriminative metrics that allow identifying vulnerable units of code.

**Keywords:** Static code metrics · Risk scores · Android
Security prediction models

## 1 Introduction

Several software systems bump into security issues during their lifetime. To ensure building highly secure software systems, developers need to invest the right amount of time in testing and debugging security issues. Nonetheless, due to limited resources, it is usually not possible to thoroughly check every file in a software system. Prioritization and different level of focus are needed to check and test more parts of the system that are more prone to be vulnerable. Regrettably, identifying these vulnerable parts is not an easy task since there are many parts in a software system, and only a few of them are vulnerable.

Software security vulnerabilities are a continuous danger to software businesses and their clients. Evaluating the security of a software system requires prioritizing resources and minimizing risks. Several available techniques can be used to identify security vulnerabilities before releasing the software such as manual inspection, static and dynamic analysis. However, these techniques were found to be error-prone and resource-intensive activities. This research aims to improve the detection of security

© Springer International Publishing AG, part of Springer Nature 2018
M. Alenezi and B. Qureshi (Eds.): *5th International Symposium
on Data Mining Applications*, pp. 84–94, 2018.
https://doi.org/10.1007/978-3-319-78753-4_8

issues by building a predictive model that will enable engineers to focus their activities on non-secure apps. Basili et al. [1] suggested that prediction models can support software project planning, scheduling, and decision-making. They enable software teams to properly allocate needed resources to modules that are more likely to be defect-prone. Generally, security vulnerabilities are a subset of defects.

In this work, we study which software static metrics have more relation with security vulnerabilities in the source code. We hypothesize that some metrics can be employed to distinguish between vulnerable or non-vulnerable code. A dataset including 1407 Android apps with different static code metrics was analyzed. The purpose of this research is to design an empirical study that aims to find the correlations among these metrics, their impact in detecting vulnerabilities in source codes and to define set of metrics which highly contributing to the prediction of security vulnerabilities. The results of this empirical study showed a strong correlation among some of the static metrics. Moreover, by applying machine learning algorithm on the complete dataset with 21 static metrics, the ability of these metrics to predict security vulnerabilities was observed with accuracy reached 94.4% in some classes. To find out the set of metrics which have more influence in detecting the insecure codes, a feature selection algorithm was also implemented. This algorithm succeeded to choose 9 out of 21 metrics which reduced the complexity of the prediction model without affecting its accuracy. As a result, this research has mainly introduced an efficient risk score prediction model for Android applications.

The rest of paper is organized as follows: Sect. 2 presents recent related work. Section 3 introduces the empirical study setup. The experiments and their results are detailed in Sect. 4. Then a discussion section is followed. Section 6 concludes the paper and presents possible future work.

## 2 Related Work

Software security research has been going for a long period of time, several topics were discussed by researchers, including security protocols and patterns to build secure systems [2], software security testing [3], vulnerability detection [4], attack prediction [5], and intrusion detection systems [6], just to name some. This shows that building software without security vulnerabilities, despite the huge advances in software development processes, is still very difficult, if not impossible [7].

Majority of similar studies concentrated on detecting and categorizing malicious Android apps through the use of permissions [8], dynamic analysis, and machine learning techniques [9]. Rahman et al. [10] investigated how effectively static code can be used to predict security risk of Android applications. Based on 21 static code metrics of 1,407 Android applications, and using radial-based support vector machine (r-SVM), they got a precision of 0.83. Syer et al. [11] examined the relationship between files defect-proneness and platform dependence in Android apps. They found that source code files that are defect-prone have a higher dependence on the platform than defect-free files. Previous studies also investigated the undesirable effects of Android apps low-quality source code. Corral and Fronza [12] examined how market success is dependent on source code quality.

Software metrics have been widely-used to build prediction models to predict faults [13, 14]. We also believe that software metrics can be used to build prediction models to predict security vulnerabilities. Several vulnerability detection techniques and tools are implemented by both commercial and open source to detect software vulnerabilities. Generally, these techniques can be categorized broadly into three main categories: static code analysis [15], emulation of security attacks (i.e., also known as penetration testing) [16], and runtime monitoring of the system behavior [17].

While using software metrics to detect and predict security vulnerabilities is not that common, we still can find quite a few studies in the literature to show some relationships between the software internal quality attributes and its security (as an external quality attribute). The work in [18], is one of the first attempts to show that there is a strong correlation between attackability (i.e., likelihood of an attack to succeed on a software system [19]) and coupling metrics (e.g., Coupling Between Objects).

A different effort was focused on predicting software security vulnerabilities using quantitative metrics [20]. A new metric called vulnerability density (i.e., number of vulnerabilities per unit of code) was proposed to be used for comparing software systems within the same category in terms of functionality. They also investigated the possibility of predicting the number of vulnerabilities. In a different study [21], the authors tried to understand where the majority of the vulnerabilities occur in a software. They were not successful in finding a correlation between complexity or buffer usage and the number of vulnerabilities. The authors in [22] focused more on complexity metrics to predict failures and security vulnerabilities in software. They employed machine learning approach to build a model using nine complexity metrics to predict vulnerabilities. They showed that complexity metrics (e.g., Cyclomatic Complexity) can predict vulnerabilities, but with a very high false negative rate. In [23], the authors used nine function-level complexity metrics to build a framework to automatically predict vulnerabilities (in addition to other coupling and cohesion file-level metrics, in a total of 17). Several file-level metrics and development activity metrics (a total of 28) were used to distinguish between vulnerable and neutral files in [24]. The results showed their effectiveness in discriminating and predicting vulnerable files.

Recently, the authors in [25] used 27 function-level metrics to examine the correlation between the software internal quality and security vulnerabilities. Although they did not find a strong correlation between these metrics and the number of vulnerabilities, they found that software metrics can be used to discriminate vulnerable functions from non-vulnerable ones.

The following sections present the structure of an empirical study, its implementation, results' discussions and analysis. This study aims to investigate the impact of static code metrics in detecting insecure software code, which metrics and to what extent they can contribute to propose high-performance prediction models.

## 3    Empirical Study Setup

This section presents the empirical study conducted to examine the impact of static code metrics in predicting the level of vulnerability in Android apps. Figure 1 shows the empirical study structure.

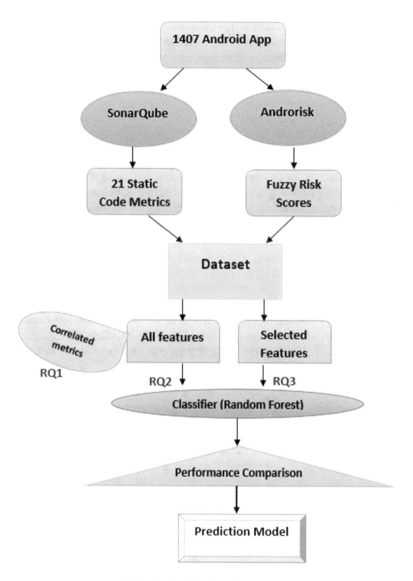

**Fig. 1.** Empirical study structure

This study has used a dataset of 1407 Android apps obtained from the authors of the work presented in [10]. SonarQube [26] analyzed these apps to extract 21 static code metrics. SonarQube is a well-known tool that uses source code static analysis and produces metrics. Since Android applications are built using Java, SonarQube was used to extract the static code metrics from the Android Java source code files. The 21 static code metrics are classified as follows:

- **Object-oriented:** Class complexity, Comment lines, Complexity, Density of comment lines, Files, File complexity, Function complexity, Lines, Lines of code, Methods, Number of classes, Percentage of comments, Percentage of duplicated lines. The
- **Bad Coding Practice:** Blocker practices, Critical practices, Major practices, Minor practices, Total bad coding practices
- **Duplication:** Duplicated blocks, Duplicated files, Duplicated lines

Moreover, Androrisk [27] tool was used to give a risk score to the Android applications using fuzzy logic. This risk score is an approximation of the amount of security and privacy risk for the Android application. Androrisk calculates a risk score between 0 and 100 for each app based on different permissions and settings used by the application. Each permission has a weight depending on its sensitivity and risk (i.e. access to the Internet, SMS messages, or payment systems). The presence of more dangerous functionality in the app (i.e. a shared library, use of cryptographic functions, the reflection API) also has an impact on the risk score. Androrisk reports the security risk score for each application. Androrisk is freely available, open-source, and has the ability to quickly process a large number of apps.

The complete Dataset used by this research's experiments includes the values of the 21 static metrics for each one of the 1407 apps. Also, the dataset categorizes the risk scores into No, Low, Medium, and High based on their statistical distribution to ease the prediction. Table 1 describes the classes/labels and their frequencies in the dataset.

**Table 1.** Classes categories in the dataset

| Class | Number |
|--------|--------|
| High | 767 |
| Low | 28 |
| Medium | 22 |
| No | 590 |
| Total | 1407 |

Using this dataset, we designed our experiments to answer the following research questions:

- RQ1: Is there any correlations among the static code metrics?
- RQ2: Are static code metrics able to predict security vulnerabilities?
- RQ3: Can static code metrics be ranked according to their contributions in predicting security vulnerabilities?

The following section illustrates how the conducted empirical study answers these research questions.

# 4   Experiments and Results

To answer the first research question (RQ1), Spearman's Rank Correlation Coefficient was performed. Spearman's correlation can be used to evaluate the statistical dependency between two metrics [28]. One of the strongest points of Spearman is the fact that it does not need a normal distribution of data. Table 2 presents the results obtained from the analysis. Two variables are considered strongly correlated if the value is higher than 0.9 [28]. Since the dataset contains static code metrics, some of these metrics might have similar definitions that could cause redundancy.

**Table 2.** Strongly correlated static code metrics

|            | LOC  | Functions | Complexity | Files | Violations | Lines |
|------------|------|-----------|------------|-------|------------|-------|
| Classes    | 0.96 | 0.97      | 0.97       | 0.93  |            | 0.95  |
| LOC        |      | 0.98      | 0.99       | 0.94  | 0.93       | 0.99  |
| Functions  |      |           | 0.98       | 0.93  | 0.9        | 0.98  |
| Complexity |      |           |            | 0.92  | 0.92       | 0.97  |

To answer the second research question (RQ2), a prediction model can be built on the available dataset. The problem is formulated as a classification problem where the static metrics are considered features and the risk level is considered class label.

A classic software security prediction model is trained using static code metrics and security data that have been collected from already developed software. After creating the prediction model, it can be used in new programs that do not have security information.

Based on previous studies [24, 29], we chose Random Forest as our machine learning algorithm to build this prediction model. Random Forests is an ensemble learning method, which produces several decision trees at training time. Each tree gives a class label. The Random Forests classifier selects the class label that has the mode of the classes output by individual trees. The algorithm combines several decision tree classifiers, each one fitted on a random sub-sample of a dataset, making it more accurate and robust to outliers and noise than a single classifier. The Random Forest classifier was implemented in WEKA tool with default parameter settings specified in WEKA.

To get the classification results, we performed a 10-fold cross-validation, which randomly partitions the data into 10 folds, with each fold being the held-out test fold exactly once. The metrics used to evaluate the performance of the classifier are namely Precision, Recall, and ROC (Receiver Operating Characteristic). The authors of [30] argue that ROC is the best measure to report the classification accuracy. In the collected data set, the number of apps with security problems is much lower than the number of apps with no security problems. The Area Under the ROC Curve (AUC) is a preferred measure since it considers the ability of a classifier to differentiate between the two classes. AUC is sound more than other performance metrics since it has lower variance. The AUC value ranges from 0 to 1. The ROC curve characterizes the trade-off between true positives and false positives. The goodness of a classifier is judged based on how large an area under the curve is. Precision measures how many of the vulnerable instances returned by a model are actually vulnerable. The higher the precision is, the fewer false positives exist. Recall measures how many of the vulnerable instances are actually returned by a model. The higher the recall is the fewer false negatives exist.

Table 3 shows the performance of the random forest-based prediction model, where Table 4 displays the confusion matrix in case of considering all features (21 static code metrics). Overall, the best prediction is observed in case of High and No classes.

**Table 3.** Random forest classification with all features

| Precision | Recall | ROC Area | Class |
|-----------|--------|----------|-------|
| 0.854 | 0.846 | 0.944 | No |
| 0.941 | 0.571 | 0.786 | Low |
| 0.909 | 0.455 | 0.870 | Medium |
| 0.867 | 0.898 | 0.938 | High |
| **0.864** | **0.863** | **0.936** | Weighted avg. |

**Table 4.** Confusion matrix for all features

| High | No | Low | Medium | |
|------|-----|-----|--------|---|
| 689 | 76 | 1 | 1 | **High** |
| 91 | 499 | 0 | 0 | **No** |
| 6 | 6 | 16 | 0 | **Low** |
| 9 | 3 | 0 | 10 | **Medium** |

Looking at the problem from a practitioner's point of view, it is preferred to have a model with a small set of metrics. Usually, some of these metrics provide redundant or no new knowledge. In this work and to answer the third research question (RQ3), we did a feature selection to find the best subset of the 21 metrics and to rank them with regard to their contribution in predicting security vulnerabilities in order to build a reliable prediction model.

Hall and Holmes [31] categorized feature selection algorithms to (1) algorithms that evaluate individual attributes and (2) algorithms that evaluate a subset of attributes. Since we are interested to find the best subset of features, we used the second category. Correlation-based feature selection (CFS) is an automatic filter algorithm that does not need user-defined parameters. Features selection implies removing irrelevant and redundant features. CFS calculates a heuristic measure of the merit of a feature subset from pair-wise feature correlations and a formula adapted from test theory. The subset with the highest merit found during the search is reported. After running CFS, the selected features are classes, Density of comment lines, files, directories, File complexity, violations, duplicated blocks, lines, and critical violations.

Table 5 shows the performance of the prediction model after implementing the features selection which reduces around 43% of the features. The results reveal maintaining high accuracy in comparison to the model built based on all features. Similar observations regarding the confusion matrices as can be seen in Table 6.

It is noteworthy knowing the most influential static code metrics that contribute to predicting security vulnerabilities. The most influential metrics can be computed using gain ratio [31]. The gain ratio provides a normalized measure of the contribution of each feature to the classification. Table 7 reports the rank of the selected metrics. The higher the gain ratio, the more important the feature to predict security vulnerabilities.

**Table 5.** Random forest classification after feature selection

| Precision | Recall | ROC Area | Class |
|---|---|---|---|
| 0.844 | 0.846 | 0.941 | **No** |
| 0.941 | 0.571 | 0.802 | **Low** |
| 0.909 | 0.455 | 0.879 | **Medium** |
| 0.862 | 0.885 | 0.934 | **High** |
| **0.857** | **0.856** | **0.934** | **Weighted avg.** |

**Table 6.** Confusion matrix after features selection

| High | No | Low | Medium | |
|---|---|---|---|---|
| 679 | 86 | 1 | 1 | **High** |
| 91 | 499 | 0 | 0 | **No** |
| 8 | 4 | 16 | 0 | **Low** |
| 10 | 2 | 0 | 10 | **Medium** |

**Table 7.** Features with high gain ratio values

| Metric | Gain ratio |
|---|---|
| comment_lines_density | 0.1078 |
| Lines | 0.0973 |
| Files | 0.0839 |
| Violations | 0.0818 |
| duplicated_blocks | 0.0787 |
| Classes | 0.0763 |
| Directories | 0.071 |
| critical_violations | 0.0701 |
| file_complexity | 0.0318 |

## 5   Discussion

All the metrics used in this study are static which means developers and organizations can calculate them very easily. The features selected by the feature selection methodology are classes, Density of comment lines, files, directories, File complexity, violations, duplicated blocks, lines, and critical violations. The number of classes determines the size of the app, which plays an important role in determining the security of the app. More classes mean more attack surface. For Density of comment lines, it is very common in the software engineering community that if you have a very complex code, developers tend to write more comments to explain it. Files, directories, and lines have the same philosophy of the classes. For the file complexity, complexity is always the enemy of security [32]. Regarding duplicated blocks and critical violations, they are patterns found to be unhealthy for software systems especially security and privacy.

# 6 Conclusion

This papers presented an empirical study to examine the impact of static code metrics in predicting security vulnerabilities in Android applications. Three main research questions have been raised and answered in this research. The study results showed different levels of correlation among the code metrics. A direct influence for the code metrics in predicting security vulnerabilities was also observed. Moreover, these metrics could be ranked according to their contributions to providing high prediction rate. Feature selection algorithm played an important role in selecting the most influential metrics. The algorithm reduced around 43% of the static metrics which resulted in producing a lightweight, reliable prediction model.

As for future work, other static metrics could be considered, other tools with different ranking philosophies could be examined to enhance the ranking system and introduce prediction models with even higher accuracy.

# References

1. Basili, V.R., Briand, L.C., Melo, W.L.: A validation of object-oriented design metrics as quality indicators. IEEE Trans. Softw. Eng. **22**(10), 751–761 (1996)
2. Fernandez-Buglioni, E.: Security Patterns in Practice: Designing Secure Architectures Using Software Patterns. Wiley, Chichester (2013)
3. Wysopal, C., Nelson, L., Dustin, E., Zovi, D.D.: The Art of Software Security Testing: Identifying Software Security Flaws. Pearson Education (2006)
4. Antunes, N., Vieira, M.: Benchmarking vulnerability detection tools for web services. In: 2010 IEEE International Conference on Web Services (ICWS), pp. 203–210. IEEE (2010)
5. Abdlhamed, M., Kifayat, K., Shi, Q., Hurst, W.: Intrusion prediction systems. In: Information Fusion for Cyber-Security Analytics, pp. 155–174. Springer, Cham (2017)
6. Messier, R.: Intrusion detection systems. In: Network Forensics, pp. 187–209 (2017)
7. Cusumano, M.A.: Who is liable for bugs and security flaws in software? Commun. ACM **47**(3), 25–27 (2004)
8. Gorla, A., Tavecchia, I., Gross, F., Zeller, A.: Checking app behavior against app descriptions. In: Proceedings of the 36th International Conference on Software Engineering, pp. 1025–1035. ACM (2014)

9. Peng, H., Gates, C., Sarma, B., Li, N., Qi, Y., Potharaju, R., Nita-Rotaru, C., Molloy, I.: Using probabilistic generative models for ranking risks of Android apps. In: Proceedings of the 2012 ACM Conference on Computer and Communications Security, pp. 241–252. ACM (2012)

10. Rahman, A., Pradhan, P., Partho, A., Williams, L.: Predicting Android application security and privacy risk with static code metrics. In: Proceedings of the 4th International Conference on Mobile Software Engineering and Systems, pp. 149–153. IEEE Press (2017)

11. Syer, M.D., Nagappan, M., Adams, B., Hassan, A.E.: Studying the relationship between source code quality and mobile platform dependence. Softw. Qual. J. **23**(3), 485–508 (2015)

12. Corral, L., Fronza, I.: Better code for better apps: a study on source code quality and market success of Android applications. In: Proceedings of the Second ACM International Conference on Mobile Software Engineering and Systems, pp. 22–32. IEEE Press (2015)

13. Ostrand, T.J., Weyuker, E.J., Bell, R.M.: Predicting the location and number of faults in large software systems. IEEE Trans. Softw. Eng. **31**(4), 340–355 (2005)

14. Menzies, T., Greenwald, J., Frank, A.: Data mining static code attributes to learn defect predictors. IEEE Trans. Softw. Eng. **33**(1), 2–13 (2007)

15. Chess, B., McGraw, G.: Static analysis for security. IEEE Secur. Priv. **2**(6), 76–79 (2004)

16. Arkin, B., Stender, S., McGraw, G.: Software penetration testing. IEEE Secur. Priv. **3**(1), 84–87 (2005)

17. Ko, C., Ruschitzka, M., Levitt, K.: Execution monitoring of security-critical programs in distributed systems: a specification-based approach. In: IEEE Symposium on Security and Privacy, Proceedings, pp. 175–187. IEEE (1997)

18. Liu, M.Y., Traore, I.: Empirical relation between coupling and attackability in software systems: a case study on DOS. In: Proceedings of the 2006 Workshop on Programming Languages and Analysis for Security, pp. 57–64. ACM (2006)

19. Howard, M., Pincus, J., Wing, J.M.: Measuring relative attack surfaces. In: Computer Security in the 21st Century, pp. 109–137. Springer, Heidelberg (2005)

20. Alhazmi, O.H., Malaiya, Y.K., Ray, I.: Measuring, analyzing and predicting security vulnerabilities in software systems. Comput. Secur. **26**(3), 219–228 (2007)

21. Neuhaus, S., Zimmermann, T., Holler, C., Zeller, A.: Predicting vulnerable software components. In: Proceedings of the 14th ACM Conference on Computer and Communications Security, pp. 529–540. ACM (2007)

22. Shin, Y.: Exploring complexity metrics as indicators of software vulnerability. In: Proceedings of the 3rd International Doctoral Symposium on Empirical Software Engineering, Kaiserslautem, Germany (2008)

23. Chowdhury, I., Zulkernine, M.: Using complexity, coupling, and cohesion metrics as early indicators of vulnerabilities. J. Syst. Architect. **57**(3), 294–313 (2011)

24. Shin, Y., Meneely, A., Williams, L., Osborne, J.A.: Evaluating complexity, code churn, and developer activity metrics as indicators of software vulnerabilities. IEEE Trans. Softw. Eng. **37**(6), 772–787 (2011)

25. Alves, H., Fonseca, B., Antunes, N.: Software metrics and security vulnerabilities: dataset and exploratory study. In: 2016 12th European Dependable Computing Conference (EDCC), pp. 37–44. IEEE (2016)

26. Campbell, G., Papapetrou, P.P.: SonarQube in Action. Manning Publications Co., Greenwich (2013)

27. Dunham, K., Hartman, S., Quintans, M., Morales, J.A., Strazzere, T.: Android Malware and Analysis. CRC Press, Boca Raton (2014)

28. Montgomery, D.C., Runger, G.C.: Applied Statistics and Probability for Engineers. John Wiley & Sons, New York (2010)

29. Abunadi, I., Alenezi, M.: An empirical investigation of security vulnerabilities within web applications. J. UCS **22**(4), 537–551 (2016)
30. Czibula, G., Marian, Z., Czibula, I.G.: Software defect prediction using relational association rule mining. Inf. Sci. **264**, 260–278 (2014)
31. Hall, M.A., Holmes, G.: Benchmarking attribute selection techniques for discrete class data mining. IEEE Trans. Knowl. Data Eng. **15**(6), 1437–1447 (2003)
32. Lagerström, R., Baldwin, C., MacCormack, A., Sturtevant, D., Doolan, L.: Exploring the relationship between architecture coupling and software vulnerabilities. In: International Symposium on Engineering Secure Software and Systems, pp. 53–69. Springer, Heidelberg (2017)

# Toward Stream Analysis of Software Debugging Data

Sarab A. AlMuhaideb$^{(\boxtimes)}$ and Sarah M. AlMuhanna

Computer Science Department, Prince Sultan University, Riyadh, Saudi Arabia
smuhaideb@psu.edu.sa, Sarah.almuhanna@hotmail.com

**Abstract.** Data stream mining is considered one of the new generations in data mining. Analyzing big data in a fast and reliable way is essential for identifying program failures and their reasons. Widely used data mining algorithms for this purpose rely on offline data, but the new trend is toward collaborative environments, in which data stream classification can be of high value. In this paper, the properties, characteristics, and requirements of data used in debugging are identified. Then, experiments are conducted, using the Massive Online Analysis Framework for Stream Classification and Clustering (MOA), to specify the most suitable class of data stream classification algorithms in this framework. Results showed that when applying data stream classification algorithms to data streams and data sets having similar characteristics to those expected in software debugging environments, the Hoeffding algorithms group showed notable competence in reference to other data stream classification algorithms. Further, although ensemble-based methods are known to be better performing, they have not shown noteworthy added value in the case of mining debugging software engineering data. Therefore, we recommend applying the Hoeffding algorithms group since it does not require additional cost in terms of time, memory, and model size, which is required when applying the ensemble algorithms group.

**Keywords:** Data stream classification · Big data · Data mining
Software engineering · Debugging · Ensemble algorithms · MOA

## 1 Introduction

Due to the increased amount of available data produced during the software development phases, data mining techniques are required. Debugging is the most effort-consuming phase in the software development life cycle (SDLC) [1]. Analyzing large amounts of data in a fast and reliable way is essential for identifying program failures and their reasons. In the software engineering environment, most recent data mining algorithms are conducted using offline data that are collected and stored in repositories [2–4].

In modern software engineering (SE) environments, the collaborative environment is a new trend that leads to processing a massive amount of data, which will require accordingly extended time. Traditional data mining techniques have limited value in this case. In order to cope with the modern integrated software engineering environments, especially collaborative ones, new data mining algorithms are required to be

© Springer International Publishing AG, part of Springer Nature 2018
M. Alenezi and B. Qureshi (Eds.): *5th International Symposium on Data Mining Applications*, pp. 95–109, 2018.
https://doi.org/10.1007/978-3-319-78753-4_9

conducted using online data to facilitate fast processing of debugging data in real time. Data stream mining is considered one of the new generations in data mining, and can be of high value in this regard. Our aim is to investigate the effectiveness of using online stream analysis in software debugging.

In this research we are interested in the classification of data stream from big data analytics. In specific, we will emphasize data stream mining techniques which aids in solving debugging problems. The classification problem in debugging environment is analyzed, in order to identify its requirements and the characteristics of data used. Next, techniques of data stream classification algorithms that are suitable to the software debugging problem requirements and are available in open-source online libraries are studied. Accordingly, experiments are conducted to specify the most suitable class of data stream classification algorithms for debugging problems. Experiments used a benchmark of 12 data stream classification algorithms and a total number of 14 data streams generators and large data sets from open source libraries. Results showed that when applying data stream classification algorithms to data streams and data sets having similar characteristics to those expected in software debugging environment, the Hoeffding algorithms group showed notable competence in reference to other data stream classification algorithms. Further, although ensemble-based methods are known to be better performing, they have not shown noteworthy added value in the case of mining debugging software engineering data.

In this paper, first, we will go over data mining in software debugging in Sect. 2. Section 3 presents the work related to mining debugging data. The section concludes with the basic debugging data characteristics and the requirements of mining debugging data. The experimental study and its results are presented in Sect. 4. In Sect. 5, we will highlight the results of cases where the examined data streams and data sets have similar characteristics to those expected in software debugging environments. Finally, in Sect. 6 the overall conclusion of this research, limitation, and future work are stated.

## 2   Basic Concepts

The process of analyzing the execution of a certain program containing bugs is considered a data mining task [5]. Classification is data mining technique and is define as follows [6]:

Given a set of $N$ training examples in the form of $(x, y)$, where $y$ is a discrete class label and $x \in \mathbb{R}^m$ is a vector of attributes the form $\langle x_{i1}, x_{i2}, \ldots, x_{im} \rangle$. To preform classification on data, is to find $y = f(x)$ that will predict with high accuracy the class $y$ according to the features of $x$.

In case of classification mechanism in debugging phase, we are considering two classes, which are success and failure. Having a test suite $Test_{k=1}^n$, where every test case in the suit is defined as: $t_k = (inp_k, out_k)$. $inp_k$ and $out_k$ are respectively the input and the desirable output of the test case. An execution is considered successful only if the output of the program based on $t_k$ is identical to its input.

The basic stages for mining software engineering data are as follows [7]. First, SE data to be mined are explored and gathered and the task of software engineering to be applied is defined. A large volume of data is collected that comprise heterogeneous types. SE data must be preprocessed and prepared for the data mining procedure. The data mining algorithm is then developed or adopted depending on the data mining requirement concluded from the first two stages. Offline mining is applied on already stored and gathered data. The new trend is toward a collaborative environment [8] where SE data are collected and mined directly to provide immediate feedback. Applying data stream classification can be of high value. Finally, transforming the results of the data mining algorithm into the required format to help in applying SE task.

# 3  Related Work

Debugging is the process responsible of removing errors and causes of bugs that were discovered in the testing phase [9]. Hassan and Xie present a classification method for tracing non-crashing bugs using SVM classification and closed graph mining [10]. The program execution at each function level is represented as a behavior graph. From the behavior graphs, feature data sets are extracted (e.g. edges, frequent graphs, and closed frequent graphs) and a SVM algorithm with a linear kernel is applied to isolate program regions that may lead to faulty executions.

Recently, a large number of software programs provide the ability to detect runtime failures by transmitting a failure report to the developer through the internet with user permission. These reports include specific information about the failure. Due to the huge number of users of these programs, such as Visual studio, Microsoft, Explorer, etc., this will result in an enormous number of reports that may require big data mining techniques. Several algorithms were introduced to classify software failure reports using offline data mining [3, 4]. Podgurski et al. [3] applied their classification mechanism on different compilers to classify the reported failures with their causes. Francis et al. [4] used two techniques that have tree based structures for identifying and grouping failed executions with related causes. One of them uses a dendrogram, which is a tree-based diagram representation of hierarchical cluster analysis output. The other technique is a decision tree classification using the CART algorithm. These techniques are applied on execution profiles of several compilers. The experiments showed that both techniques were efficient in classifying failed and successful execution.

In addition to the previous paper, Renieres and Reiss developed a Nearest Neighbor (NN) algorithm for localizing program's bugs [11]. The algorithm relies on program spectra, which is a collection of program execution features used in localizing a program's bug.

In another study, Wong et al. [12] trained an RBF neural network on statement coverage of virtual test cases. A coverage vector defines the program statement(s) covered by each test case, and serves as input to the neural network. Based in the execution output of the program using a test case as successful or failure, the network produced output is interpreted as the suspiciousness of the corresponding covered statement to contain bugs. A following study by Park et al. [13] designed a polynomial function-based Neural Network for software module classification as faulty/fault-free using two categories of discriminant functions: linear and quadratic. The classifier comes in an "if-then" rules form. Fuzzy C-Means clustering is used to derive the premise of the rules while the consequence is obtained using the polynomial functions.

Finally, Zeller has applied the Delta Debugging (DD) algorithm [14]. DD is a separate and conquer procedure that separates the related values and variables (cause-effect chain) that caused the program to fail. This algorithm operates by performing a binary search on the memory graphs [15] of both failed and successful executions. This process systematically leads to narrowing the difference between a successful and failed execution in order to have a small group of variables that are suspected of causing the failure.

**Observations.** All of the above-reviewed papers, used test case generators for program execution. After test beds are executed, behavior graphs, dendrograms, etc. are generated optionally. Then the data are extracted to be fed to the classification algorithm in order to decide whether the execution succeeded or failed. Figure 1 shows a high-level description of the process for data mining in debugging phase.

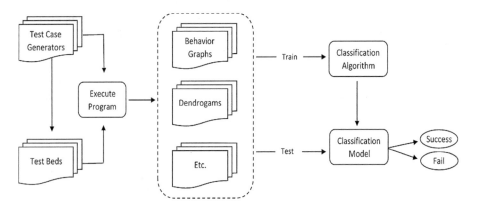

**Fig. 1.** Data mining in debugging phase.

Studying the earlier-reviewed papers, we have observed that the required characteristics of debugging environment are noise free data and binary class classification problem type. As for the requirements, data should be online, have a single pass, and the adopted algorithm should be transparent model. These observations are explained next.

- *Noise.* Han et al. defined noise as: "A random error or variance in a measured variable" [6]. Since there was no human intervention or machine failure assumed in the previous experiments, we conclude that *the data do not include noise.*
- *Classification problem type.* The experiments used test case generators for program execution. After test beds are executed, behavior graphs, dendrograms, etc. are generated optionally. Then the data are extracted to be fed to the classification algorithm in order to decide whether the execution succeeded or failed. The decision made is based on the output of the test case $t_k = (inp_k, out_k)$. Where a failure execution occurs only if the output does not match $out_k$. In other words, the classification problem used is considered a *binary class problem.*
- *Data Type.* The data used in the experiments belong to the heterogeneous data type. Input data depend on the test bed, which could be of any type: numerical,

categorical, graphs, etc. Therefore, there is no specific type of data that is required to focus on.

- *Transparent models.* In the context of software debugging, the outcome of data mining would be predicting either a failed or a successful execution. In the case of a failed execution, the reasons behind the system failed should be clear. For that purpose, transparent models are recommended because they provide reasoning along with the prediction. Notably, transparency alone is not enough for a model to be comprehensible by humans. The size of the generated model is an important factor. For example, consider models having thousands of production rules. These models are not easily managed and understood.

In the context of software debugging, the data that are used in the training phase consist of a huge amount of data and have an infinite length due to the collaborative environment. Therefore, data processing and response should be fast and online where data are not stored and require only a single pass.

## 4 Experimental Study

In this research the classification algorithms are chosen based on three criteria. First, the algorithm should require a single pass on an infinite data stream. Another requirement described in Sect. 3 is that the selected algorithm need to generate transparent models, as opposed to black-box models such as those produced by ANNs and SVMs. Third, the selected algorithms need to be available in open-source online libraries. The classifiers and data stream generators used in this research are available in Massive Online Analysis (MOA) [16], which is a framework for implementing new algorithms and executing experiments for online learning using a huge sequence of data. The adopted data stream classification algorithms (Table 1) are grouped into the following

**Table 1.**  Adapted data stream classifiers.

| Category Name | Classifiers |
|---|---|
| Hoeffding algorithms | • Hoeffding tree (HT[a]) [17]<br>• Adaptive Hoeffding Option Tree (AdaHOT) [18]<br>• Hoeffding Adaptive Tree (HAT) [19]<br>• Hoeffding Option Trees (HOT) [20] |
| Ensemble methods | • Oza Bagging (OB) [21]<br>• Oza Bagging Adwin (OBA) [18]<br>• Oza Boosting (OBST) [21]<br>• Oza Boosting Adwin (OBSTA) [18] |
| Rule classifiers | • RC, implementation of Very Fast Decision Rules VFDR [22]<br>• RCNB, implementation of Very Fast Decision Rules Naïve Bayes VFDRNB [22] |
| Weka classifiers | • J48 [23]<br>• Random Tree (RT[b]) |

[a]In MOA GUI, the implementation of Hoeffding tree algorithms is referred to as HT or VFDT.
[b]http://weka.sourceforge.net/doc.dev/weka/classifiers/trees/RandomTree.html.

four groups: (i) Hoeffding algorithms, (ii) ensemble algorithms, (iii) rule classifiers, and (iv) Weka classifiers. As for Weka classifiers that are used in MOA GUI, they learn from a chunk of the data stream that is buffered from MOA.

The data stream classification algorithms will be evaluated by the following performance measures: *speed;* measured in seconds for a single cycle; *predictive accuracy* for a single cycle which is calculated by dividing the number of correctly classified cases by the total number of classified cases; and *size* of the decision tree in terms of number of tree nodes for a single cycle. According the group of ensemble methods, it must keep in consideration that these methods report the average size of the generated trees, not the total. As for rule classifiers and Weka classifiers, they do not apply this measure.

In our experiment, we have used the prequential evaluation in testing several decision tree classifiers and rule classifiers using both data stream generators and large data sets. Since this research is interested in the characteristics of data found in software debugging environments detailed in Sect. 3, the experimental results are arranged into three groups as follows. Examining noise factor, examining number of classes factor, and examining binary class classification for data without noise.

The data sets have been chosen based on two criteria. First, they are available in big data open-source libraries. Second, they are selected such as to match the desired characteristics in terms of noise and number of classes in each group. Similarly, the settings of the data stream generators have been set accordingly. These factors are tested on several algorithms such as: Hoeffding algorithms, ensemble methods, rule classifiers, and Weka classifiers.

## 4.1    Experimental Setup

The number of instances in each data stream generator used in the experiment is set to 1,000,000 instances. The default value of the sample frequency is kept the same, and this means that we will have 10 classification cycles for each generator. The number of

**Table 2.** Properties of data stream generators in MOA.

| Data stream generator | #Att. | #Classes | Noise | Data type |
|---|---|---|---|---|
| Agrawal generator [24] | 9 | 2 | Default %5 | 6 Numerical, 3 categorical |
| Hyperplane generator [25] | $10^a$ | $2^a$ | Default %5 | Boolean |
| LED 7 generator [26] | 7 | 10 | Default %10 | Boolean |
| LED 7 generator drift | 7 | 10 | Default %10 | Boolean |
| Random RBF generator [25] | $10^a$ | $2^a$ | %0 | Numerical |
| Random RBF generator drift | $10^a$ | $2^a$ | %0 | Numerical |
| Random tree generator [17] | $10^a$ | $2^a$ | Includes noise | $5^a$ Numerical, $5^a$ nominal |
| SEA generator [27] | 3 | 2 | Default %10 | Numerical |
| STAGGER generator [28] | 3 | 2 | %0 | Boolean |
| Waveform 40 generator [26] | 40 | 3 | Optional | Numerical |
| Waveform 40 generator drift | 40 | 3 | Optional | Numerical |

[a]Default value.

instances for each data set differs from one data set to another. The average of results over the cycles is reported. For the classifiers that were adopted in the experiment, the default settings were applied. In the case of ensemble methods, the number of base learners is set to 10 and the average size of the generated trees is considered in each execution. The experiment will be conducted on a machine that contains at least a 2.89 GH processor and 3.25 GB RAM, which can be found in a personal laptop. A description of the adopted data stream generators is listed in Table 2, and data sets are presented in Table 3.

**Table 3.** Properties of the adopted classification data sets.

| Data sets | #Attributes | #Instances | Data type |
|---|---|---|---|
| Bank marketing | 16 | 45,211 | Continuous, nominal |
| Skin segmentation | 3 | 245,057 | Continuous |
| Census income | 41 | 299,285 | Integer, nominal |

## 4.2    Experimental Results

As we mentioned previously, the results of the experiments are grouped according to two factors: noise and number of classes. We will analyze the results of applying these factors as follows:

**Examining Noise Factor.** In this factor, we have tested the adopted algorithms on several data stream generators that provide the ability to adjust the noise percentage. These generators are: Agrawal, Hyperplane, Led with or without concept drift, Sea, and Waveform with or without concept drift. The classification experiment is conducted twice on each set: without noise and including noise.

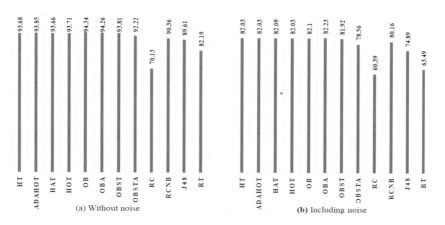

**Fig. 2.** Accuracy average results of testing MOA classifiers using data stream generators that examines noise factor.

*Experiment #1: Without Noise.* In this experiment, the noise level is set to zero for all generators. Figure 2(a) shows the average results of accuracy when learning from the data stream generator without noise. The tests failed for the RC when using the Sea and the Hyperplane generator. As for the rest, the results indicate that OB, OBA, AdaHOT, OBST, and HOT recorded the highest accuracy algorithms, while RCNB and J48 had the least accuracy. In addition, applying the Friedman test on all classifiers as well as on the Hoeffding algorithms group, combined with the ensemble methods group, has detected a significant difference among different models. However, no significant difference was detected on the top 5 highest accuracy algorithms. Secondly, the result of the smallest number of nodes were generated from the following algorithms: OBA, HT, OB, and HAT, as stated in Fig. 3(a). Finally, the evaluation time for the algorithms

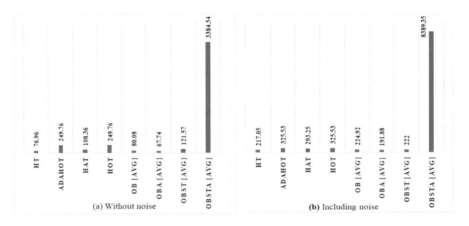

**Fig. 3.** Size average results of testing MOA classifiers using data stream generators that examines noise factor.

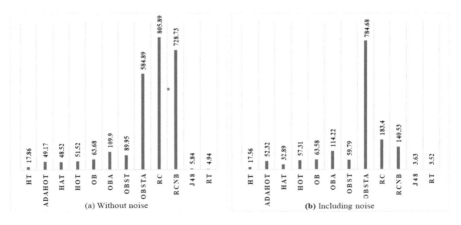

**Fig. 4.** Evaluation time average results of testing MOA classifiers using data stream generators that examines noise factor.

in Fig. 4(a) shows that the fastest performance was for RT, J48, HT, HAT, and AdaHOT, respectively.

*Experiment #2: Including Noise.* The noise level is set to a default value, which ranges from 5–10%. Algorithms with the highest accuracy were scored by OBA, OB, HAT as shown in Fig. 2(b). Furthermore, Friedman test detected a significant difference. However, no significant difference was detected when comparing the various models in the ensemble methods group against the Hoeffding algorithms group. Figure 3(b) indicates that OBA, HT, OBST, and HAT had the smallest number of nodes. For the evaluation time, Fig. 4(b) shows that RT, J48, HT, HAT, and AdaHOT had the fastest performance while the OBSTA algorithm was the slowest. Testing the rule classifiers failed on the Waveform generator and the Agrawal generator.

**Examining Number of Classes Factor.** The chosen algorithms are tested on several data stream generators that pride the ability to adjust the number of classes for each generator. These generators are: Hyperplane, Random RBF with or without concept drift, and Random Tree. Similarly, the classification experiment is conducted twice on each set: the binary class and multiple class, as follows.

*Experiment #3: Binary Class.* The number of classes is set to 2 for all generators in this experiment. Figure 5(a) revealed that OBST, OB, OBA, AdaHOT, and HOT have the highest accuracy algorithms. RCNB and J48 had the least accuracy. Additionally, there is a significant difference detected when applying the Friedman test to all classifiers and to test both the ensemble methods group and the Hoeffding algorithms group. On the other hand, when performing the Wilcoxon signed rank test on every combination from the Hoeffding algorithms group and the ensemble methods group, there were no significant differences detected. As for the size of the generated algorithms, as in Fig. 6(a), OBA, HT, OB, OBST, and HAT had the smallest number of nodes. Figure 7(a) states the evaluation time learning on binary generators. The result recorded that the fastest performance was for RT, J48, HT, HOT, and AdaHOT, respectively.

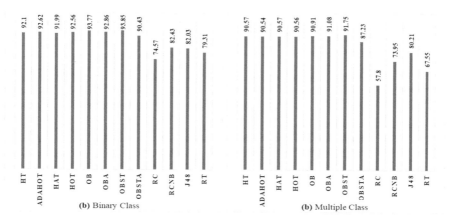

**Fig. 5.** Accuracy average results of testing MOA classifiers using data stream generators that examines number of classes factor.

**Fig. 6.** Size average results of testing MOA classifiers using data stream generators that examines number of classes factor.

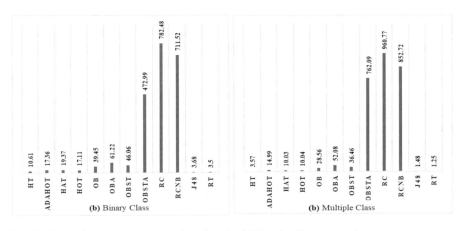

**Fig. 7.** Evaluation time average results of testing MOA classifiers using data stream generators that examines number of classes factor.

*Experiment #4: Multiple Classes.* The number of classes is set to 6 classes for all generators. Learning from the Random RBF generator with or without the concept drift, using the rule classifiers group, led to failed tests. Figure 5(b) shows algorithms with the highest accuracy which were scored by OBST, OBA, and OB, and the fourth highest accuracy was for HT and HAT, followed by HOT. Moreover, no significant difference among various models was detected when applying the Friedman test to all classifiers. OBA, HT, OB, AdaHOT, and HOT, followed by OBST, contain the smallest number of nodes, as shown in Fig. 6(b). Finally, the evaluation time in Fig. 7 (b) indicates that RT, J48, HT, HAT, and HOT have the fastest performance, while the RC algorithm has the slowest.

*Experiment #5: Binary Class Without Noise.* In this experiment, the noise level is set to zero for all generators, and the number of classes is set to 2. As for the data sets, they all feature binary classification and don't contain noise. Similar to the experiments without noise, this experiment experienced failed tests. Figure 8 reveals the results of accuracy which indicate that OBA, HAT, OBST, OBSTA, and OB recorded the highest accuracy algorithms. Yet RCNB and RC were the least accurate.

Further, there were no significant differences detected when applying the Friedman test on the top 5 highest accuracy algorithms as well as on the top 2 highest accuracy algorithms from both the ensemble methods group and Hoeffding algorithms group. Secondly, the results of the smallest number of nodes were generated from the

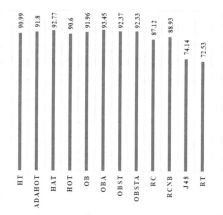

**Fig. 8.** Accuracy average results of testing MOA classifiers using binary class data stream generators and datasets without noise.

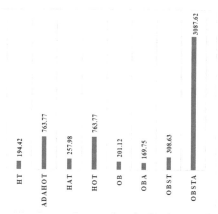

**Fig. 9.** Size average results of testing MOA classifiers using binary class data stream generators and datasets without noise.

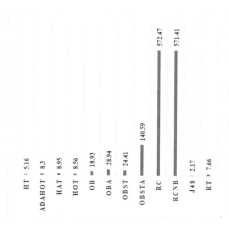

**Fig. 10.** Evaluation time average results of testing MOA classifiers using binary class data stream generators and datasets without noise.

following algorithms: OBA, HT, OB, and HAT, as stated in Fig. 9. Finally, the evaluation time for the algorithms was recorded. Figure 10 shows that the fastest performance was for J48, HT, RT, AdaHOT, and HOT, respectively.

## 5  Discussion

Since we are interested in examining data streams and data sets that have similar characteristics to those expected in software debugging environments, we will discuss the following results according to the specified cases.

**Absence of Noise.** Table 4 shows the average results of each algorithm group according to noise absence. Notably the Hoeffding algorithms group was better than the ensemble methods group. In addition, the Friedman test revealed that the difference between these groups was statistically significant. The ensemble methods group was required to perform 170.34 s more on average than the HT algorithms group when the data stream did not contain noise. The time needed by an ensemble method is proportional to the number of base learners used. The difference in model size is considerable among these groups. The model size of the ensemble methods is the average of the whole ensemble, which we set to 10. To make it clear, the actual size of the model in the case of the ensemble methods group is multiplied by 10. Therefore, the HT algorithms group has a better fit than the ensemble methods group for SE debugging tasks.

**Table 4.** Average results of applying data stream generators without noise.

|                        | HT      | Ensemble                      | Rule    | Weka   |
|------------------------|---------|-------------------------------|---------|--------|
| Accuracy               | 93.73%  | 93.66%                        | 80.25%  | 85.9%  |
| Evaluation time (sec)  | 41.77   | 212.11                        | 767.22  | 5.39   |
| Model size (nodes)     | 171.21  | $913.48 \times 10 = 9134.8$   | –       | –      |

**Binary Class.** The average results of each algorithm group are shown in Table 5. The ensemble methods group performed slightly better than the Hoeffding algorithms group. The difference was not statistically significant between AdaHOT and OBST, the best performing algorithms from each group. Additionally, the cost of ensemble methods in terms of time, memory, and the resulting model size, is much higher than Hoeffding algorithms. The additional cost required to apply ensemble methods over Hoeffding algorithms is considerable, while the gain in accuracy can be considered marginal (0.41%).

**Table 5.** Average results of applying binary class data stream generators.

|                        | HT      | Ensemble                         | Rule    | Weka    |
|------------------------|---------|----------------------------------|---------|---------|
| Accuracy               | 92.32%  | 92.73%                           | 78.5%   | 80.67%  |
| Evaluation time (sec)  | 16.11   | 154.93                           | 747     | 3.59    |
| Model size (nodes)     | 833.54  | $1952.66 \times 10 = 19526.6$    | –       | –       |

**Binary Class Without Noise.** Table 6 shows the average results of each group. Although the ensemble methods group performed slightly better than the HT algorithms group, there were no significant differences between the top 5 highest ranking accuracy algorithms as well as the highest ranking accuracy algorithms from both the ensemble methods group and the Hoeffding algorithms group. The difference between the average accuracy of the ensemble methods group and the HT algorithms group was less than 1%. The ensemble methods group was required to perform 45.48 s more on average than the HT algorithms group. Furthermore, according to the model size, in the HT algorithms group, the average size of the trees was considerably smaller.

**Table 6.** Average results of applying binary classification without noise.

|  | HT | Ensemble | Rule | Weka |
|---|---|---|---|---|
| Accuracy | 91.54% | 92.53% | 88.03% | 73.34% |
| Evaluation time | 7.74 s | 53.22 s | 571.94 s | 4.9 s |
| Model size | 494.99 nodes | 941.78 nodes × 10 = 9417.8 nodes | – | – |

As a result, the ensemble methods group will not be a beneficial option to apply in SE debugging tasks. After focusing on the evaluation time for ensemble group methods, we realize that it will require a long time to perform. This will result in affecting the online feature of processing or response time. As for the other algorithm groups, the rule classifier group was the worst performing group in both without noise experiments and the binary class experiments. Weka classifiers group was the worst performing in the binary class without noise experiments.

The research faced some limitations as follows: the data stream generators and sets that were adopted in each experiment had to adhere to the specific characteristics required in each case. As a result, there were few data stream generators and sets. In addition, the number of data stream classification algorithms that supports the requirements identified for software debugging tasks is limited. Finally, the classification tasks that were applied in debugging fields in the reviewed papers were not conducted in a collaborative environment. As far as we are concerned, we could not find published work that covers this area.

## 6   Conclusion

In this paper, we have observed that the required characteristics of the debugging environment are binary class classification and noise-free data. As for the requirements, data should be online, the adopted algorithm should allow single pass, and the model should be transparent and comprehensible in terms of size.

Additionally, in the context of the software debugging classification using MOA, the study has shown that ensemble-based methods did not provide noteworthy added value and generated models that are difficult to comprehend. On the other hand, non-ensemble based learners like Hoeffding algorithms and their variants provide fast classification results with high accuracy and manageable models that can efficiently and effectively guide the identification of failure and its causes in programs.

Data stream classification in a collaborative debugging environment is still considered a new area that is not well covered and it deserves to receive more attention. Open research areas include examining the application of data stream classification algorithms in real collaborative environments for software debugging and examining more transparent data stream classification algorithms and investigating their utility.

**Acknowledgements.** We would like to thank the reviewers for their valuable comments that improved the quality of the manuscript.

# References

1. Pressman, R.: Software Engineering: A Practitioner's Approach, 7th edn. McGraw-Hill, New York (2009)
2. Halkidi, M., Spinellis, D., Tsatsaronis, G., Vazirgiannis, M.: Data mining in software engineering. Intelligent Data Anal. **15**(3), 413–441 (2011)
3. Podgurski, A., Leon, D., Francis, P., Masri, W., Minch, M., Sun J., Wang, B.: Automated support for classifying software failure reports. In: 25th International Proceedings on Software Engineering, pp. 465–475. IEEE, Portland (2003)
4. Francis, P., Leon, D., Minch, M., Podgurski, A.: Tree-based methods for classifying software failures. In: 15th International Symposium on Software Reliability Engineering, pp. 451–462 IEEE, Saint-Malo (2004)
5. Liu, C., Yan, X., Han, J.: Mining control flow abnormality for logic error isolation. In: 6th International Proceedings of SIAM International Conference on Data Mining, pp. 106–117. Society for Industrial and Applied Mathematics, Bethesda (2006)
6. Han, J., Kamber, M., Pei, J.: Data Mining: Concepts and Techniques, 3rd edn. Elsevier, Waltham (2011)
7. Xie, T., Thummalapenta, S., Lo, D., Liu, C.: Data mining for software engineering. IEEE Comput. **42**(8), 35–42 (2009)
8. Stephens, R.: Implementing client-support for collaborative spaces. In: Zemliansky, P., St. Amant, K. (eds.) Handbook of Research on Virtual Workplaces and the New Nature of Business Practices 2008, pp. 582–594, 1st edn. IGI Global, New York (2008)
9. Zeller, A.: Why Programs Fail: A Guide to Systematic Debugging, 2nd edn. Elsevier, Burlington (2009)
10. Liu, C., Yan, X., Yu, H., Han, J., Yu, P.: Mining behavior graphs for "Backtrace" of noncrashing bugs. In: 5th International Proceedings of the SIAM Conference on Data Mining, pp. 286–297. Society for Industrial and Applied Mathematics, Newport Beach (2005)
11. Renieres, M., Reiss, S.: Fault localization with nearest neighbor queries. In: 18th International Proceedings of IEEE Conference on Automated Software Engineering, pp. 30–39. IEEE, Montreal (2003)
12. Wong, W., Debroy, V., Golden, R., Xu, X., Thuraisingham, B.: Effective software fault localization using an RBF neural network. IEEE Trans. Reliability **61**(1), 149–169 (2012)
13. Park, B.-J., Oh, S.-K., Pedrycz, W.: The design of polynomial function-based neural network predictors for detection of software defects. Inf. Sci. **229**, 40–57 (2013)

14. Zeller, A.: Isolating cause-effect chains from computer programs. In: 10th International Proceedings of ACM SIGSOFT Symposium on Foundations of Software Engineering, pp. 1–10. ACM, Charleston (2002)
15. Zimmermann, T., Zeller, A.: Visualizing memory graphs. In: Diehl, S. (ed.) Software Visualization 2002. LNCS, vol. 2269, pp. 191–204. Springer, Heidelberg (2002)
16. Bifet, A., Holmes, G., Pfahringer, B., Kranen, P., Kremer, H., Jansen T., Seidl, T.: MOA: massive online analysis a framework for stream classification and clustering. J. Mach. Learn. Res. (JMLR) 11, 44–50 (2010). Workshop and Conference Proceedings on Applications of Pattern Analysis
17. Domingos, P., Hulten, G.: Mining high-speed data streams. In: 6th International Proceedings of ACM SIGKDD International Conference on Knowledge Discovery and Data Mining, pp. 71–80. ACM, Boston (2000)
18. Bifet, A., Holmes, G., Pfahringer, B., Kirkby, R., Gavaldá, R.: New ensemble methods for evolving data streams. In: 15th International Proceedings of ACM SIGKDD International Conference on Knowledge Discovery and Data Mining, pp. 139–148. ACM, Paris (2009)
19. Bifet, A., Gavaldà, R.: Adaptive learning from evolving data streams. In: Adams, N., Robardet, C., Siebes, A., Boulicaut, J.F. (eds.) Advances in Intelligent Data Analysis VIII 2009. LNCS, vol. 5772, pp. 249–260. Springer, Heidelberg (2009)
20. Pfahringer, B., Holmes, G., Kirkby, R.: New options for Hoeffding trees. In: International Proceedings of the Australasian Joint Conference on Artificial Intelligence, pp. 90–99. Springer, Heidelberg (2007)
21. Oza, N., Stuart, R.: Online bagging and boosting. In: 8th International Workshop on Artificial Intelligence and Statistics, pp. 105–112. Morgan Kaufmann, Florida (2001)
22. Gama, J., Kosina, P.: Learning decision rules from data streams. In: 22nd International Proceedings of the Joint Conference on Artificial Intelligence, pp. 1255–1260. AAAI Press, Menlo Park (2011)
23. Quinlan, J.: C4.5: Programs for Machine Learning. 1st edn. Morgan Kaufmann, San Francisco (1993)
24. Agrawal, R., Imielinski, T., Swami, A.: Database mining: a performance perspective. IEEE Trans. Knowl. Data Eng. 5(6), 914–925 (1993)
25. Hulten, G., Spencer, L., Domingos, P.: Mining time-changing data streams. In: 7th International Proceedings of ACM SIGKDD International Conference on Knowledge Discovery and Data Mining, pp. 97–106. ACM, San Francisco (2001)
26. Breiman, L., Friedman, J., Olshen, R., Stone, C.: Classification and Regression Trees, 1st edn. Wadsworth, Belmont (1984)
27. Street, W., Kim, Y.: A streaming ensemble algorithm (SEA) for large-scale classification. In: 7th International Proceedings of ACM SIGKDD International Conference on Knowledge Discovery and Data Mining, pp. 377–382. ACM, San Francisco (2001)
28. Schlimmer, J., Granger, R.: Incremental learning from noisy data. Mach. Learn. 1(3), 317–354 (1986)

# Social Networks (SocialNets)

# Evaluating the Influence of Twitter on the Saudi Arabian Stock Market Indicators

Mohammed Alshahrani[1,2], Fuxi Zhu[1(✉)], Ahmed Sameh[3],
Lin Zheng[1], and Summaya Mumtaz[4]

[1] Computer School, Wuhan University, Wuhan, Hubei, China
fxzhu@whu.edu.cn
[2] College of Computer Science and IT, Albaha University, Albaha, Saudi Arabia
[3] Computer College, Prince Sultan University, Riyadh, Saudi Arabia
[4] Department of Informatics, University of Oslo, Oslo, Norway

**Abstract.** Investors critically analyze past pricing history, which influences their future investment decisions. Social media and news items have a significant impact on stock market indices. In this paper, we apply machine learning and NLP principles to find the correlations between Arabic sentiments and trends in the Saudi Arabian stock market, TADAWUL. More than 277K Arabic tweets were crawled and 114K tweets were annotated manually. Three types of correlations were implemented, Pearson's correlation coefficient, Kendall rank correlation and Spearman's rank correlation. Moreover, the paper illustrates that the most influential users could be predictable in the future, who can have a significant impact on the stock market trends. The first achievement of this study is the collection of the largest Arabic tweets dataset specialized in finance, which will be available to the public as soon as the annotation process is finished. The second achievement is that this is the first paper to study the influence of Twitter on the Saudi stock market using different types of correlation coefficients and investigated the role of mentions on the market trends.

**Keywords:** Twitter · Stock market · Sentiment analysis · Correlations

## 1 Introduction

This paper is an investigation related to multi-disciplinary research fields on online social networks (Maia et al. [1]; Mislove et al. [2]; Bao et al. [3]), stock markets (Chen et al. [4]; Oliveira et al. [5]; (Walter) Wang [6]), and more specifically, Twitter (Qasem et al. [7]; Cha et al. [8]) and Arabic Tweets for subjectivity and sentiment analysis (Abdul-Mageed et al. [9]; Al-Sabbagh et al. [10]; Refaee et al. [11]). Many researchers in the NLP community have investigated the sentiment analysis field. Some studies have emphasis on product reviews such as Pang et al. [12], Aciar et al. [13] and Bao et al. [3].

Kang et al. [14] focused on product reviews, as the writing is well organized and related to a specific topic. However, Tweets have different forms of writing and consist of noises, such as hashtags, URLs, and emoticons, and they even have a character limit of 140. This makes Twitter sentiment analysis extremely challenging (Bao et al. [3]).

© Springer International Publishing AG, part of Springer Nature 2018
M. Alenezi and B. Qureshi (Eds.): *5th International Symposium on Data Mining Applications*, pp. 113–132, 2018.
https://doi.org/10.1007/978-3-319-78753-4_10

In the last decade, researchers have investigated the role and influence of Twitter upon different sectors, such as public opinion, healthcare, political elections and stock markets. It is an enormous challenge to understand the behavior and trends accurately (Atsalakis et al. [15]; Silva et al. [16]; and Girijia et al. [17]).

The research community have addressed this drawback aiming to forecast the trends and predicting stock market prices (Fama et al. [18]). However, with the rapid development of new social communication technologies, especially social networks such as Twitter and Facebook, new research opportunities have appeared because of social networks generate a large amount of data in many circumstances, illustrate trends, behavior and opinions of social networks users in various fields such as economy and financial markets.

In this study, we built a Twitter API, and then crawled 277K Arabic tweets related to the stock market. During the annotation process, 114K tweets have been interpreted, which are used in the experiment. Three types of correlations were implemented, Pearson's correlation coefficient, Kendall rank correlation and Spearman's rank correlation. Furthermore, we take the variable mentions role in our consideration to identify those twitter accounts, whose tweet contributed towards market trends.

The paper is organized as follows:

1. Related work in the areas of Arabic Sentiment Analysis, Saudi Stock Market and Twitter.
2. Corpus collection divided in Data Pre-processing, Data Analysis, Data Pre-processing for stock market trend identification, Methodology for Comparing User Influence with Stock Market Prices.
3. Results of experiment.
4. Conclusion and acknowledgment.

## 2    Related Work

### 2.1    Arabic Sentiment Analysis

Researchers classify Arabic language into three categories: Classical Arabic (CA), Modern Standard Arabic (MSA) and Dialectal Arabic (DA), with respect to its morphology, syntax and lexical combinations. Dialectal Arabic is divided into varieties of Arabic dialects according to the region it belongs to, i.e. Gulf Arabic and Egyptian Arabic (Al-Sabbagh and Girju [10]).

There are many drawbacks and challenges for NLP researchers who deal with DA, because its lack of standardization, and the written free-text illustrates different forms of Arabic from MSA.

Generally Arabic users generate informal text on social networks, and bi/multi-lingual users seem to employ a mixture of languages.

Sentiment Analysis in Twitter is a difficult task, due to the complexity and variability task of sentiment parts that can be included in a single tweet of only up to 140 characters, however, it can contain a significant amount of compressed information in short form. Moreover, tweets may include sarcasm, mixed polarity or unclear sentiments, and hidden meaning contents.

Recently, Machine Learning techniques have been broadly applied in Subjectivity for Sentiment Analysis (SSA), even annotated corpora is required more than grammar or lexicon-based approaches. In addition, there are growing demands in the Arabic NLP research community to construct Arabic corpora by crawling a variety of Arabic texts in the web (Al-Sabbagh and Girju [10]; Abdul-Mageed and Diab [9]).

Korayem et al. [19] described several Arabic corpuses for Subjectivity and Sentiment Analysis, such as Abdul-Mageed and Diab's [9] AWATIF multi-genre, multi-dialect Arabic corpus, crawled from the Penn Arabic Treebank (PATB), conversations from web forums and Wikipedia user talk pages. For annotations, researchers used untrained annotators via crowd sourcing techniques to tag labels (positive, negative or neutral) to sentences, and then used linguistic trained annotators to label each sentence. The researchers built adjective polarity lexicon manually provides coarse labels to Arabic adjective.

Rushdi-Saleh et al. [20] created Opinion Corpus for Arabic (OCA), which is derived from movie review websites as a corpus of texts, there is also another English version corpus named EVOCA. OCA contains 250 positive reviews and the same number of negative reviews. The researchers manually filtered noise, such as Arabizi, multi-language reviews, non-related comments and differing spellings of proper names, from the corpus.

Elaranoty et al. [21] proposed MPQA subjective lexicon & Arabic opinion holder corpus, which is an Arabic news corpus for subjective lexicon, and contains strong and weak subjective clues due to the use of different annotations and the manual lexicon translation.

Elhawary and Elfeky [22] presented Arabic Lexicon for Business Reviews. The researchers built Arabic lexicon using the similarity graph, which is a type of graph that represents two words or phrases by nodes and the similarity on polarity or meaning are the edges between nodes. The degree of similarity between two nodes is represented by the weight of the edge. Based on an unsupervised method, the graph is regularly constructed on lexical co-occurrence from large Web corpora. To build the Arabic similarity graph, a list of seeds for positive, negative and neutral must be implemented, because the researchers applied a small set of seeds initially then performed label propagation on the Arabic similarity graph (Korayem et al. [19]).

El-Halees [23] proposed subjectivity lexicon, that was constructed manually, derived from two sources, the SentiStrength project and an online directory. The English list from the SentiStrength project was translated into Arabic, then a filtering step was manually implemented. The researcher added common Arabic words to the subjectivity lexicon.

A sentiment framework was developed by Duwairi et al. [27] to analyze Tweets, whether positive, negative or neutral sentiments. This framework can be deployed in a variety of applications, from politics to marketing. The advantages of this framework are that it can handle Arabic dialects, Arabizi and emoticons. Crowd sourcing was operated to gather a large dataset of tweets.

Ibrahim et al. [24] proposed a multi-genre labeled corpus of Modern Standard Arabic (MSA) and Arabic Dialects. The annotated corpus, MIKA, was manually crawled and tagged at sentence level with semantic orientation. The corpus has multi linguistic features, such as, negation handling, contextual intensifiers, contextual shifter, syntactic features for phrases conflicts and annotation process features.

Refaea et al. [25] presented the first available general Twitter Arabic corpus as far as it is known, which was crawled and annotated for subjectivity and sentiment analysis.

ASTD is an Arabic social sentiment analysis dataset crawled from Twitter, developed by Nabil et al. [26]. The dataset contains about 10,000 tweets categorized as objective, subjective positive, subjective negative, and subjective mixed. The researchers used standard partitioning of the dataset during the experiments to illustrate the properties and the statistics of ASTD.

## 2.2    Saudi Stock Market

In the finance sector, sentiment analysis is widely used to assist investors to follow targeted companies and examine their sentiment data in real time. Nowadays, decision makers in business have better opportunities to acquire business news and trends by sentiment analysis (Duwairi et al. [27]).

Huang et al. [28], demonstrated that historical financial data can generate financial trend predictions by building frameworks, applications and models to predict future financial data. Researchers in the community have presented different models and various data has been collected and tested. Most of the research in the field concentrate on data selection type and model selection type.

Researchers like Idvall and Jonsson [29], Li et al. [30], Pidan and El-Yaniv [31], and Zhang [32] worked in data selection aspect using the history of financial data only to predict financial trends, while others merged the history of financial data with some further indicators, such as Twitter mood, to also make predictions such as Si et al. [35], Bollen et al. [33], Mittal et al. [34], and Sprenger et al. [36]. In terms of the model selected by the various researchers, Bollen et al. [33] selected the non-linear Self-Organizing Fusion Neural Networks (SOFNN) model, Si et al. [35] used the linear Vector Auto regression (VAR) model, although Idvall and Jonsson [29], Pidan and El-Yaniv [31], and Zhang [32] presented HMM based methods. Since the forecasting can be achieved by considering the combination of historical data trends and Twitter moods (Wu et al. [37]; Yang et al. [38]) regarding modelling multiple time series that may correlate with each other, and about using multiple time series together to make predictions, are also highly related. The closest works to this current study are those by Bollen et al. [33], Pidan and El-Yaniv [31] and Huang et al. [28]. One of the earlier studies to analyze Arabic/Chinese sentiments about financial news was conducted by Ahmad et al. [39]. Local grammar approach was used in their study to extract Arabic/Chinese sentimental phrases. Tests conducted by these researchers showed that the accuracy of sentiment extraction was in the range of 60–75%. Almas and Ahmad [40] investigated the financial news by extracting English/Arabic/Urdu sentiments. The outcomes of their research illustrate F-measure improve the identification of the polarity of Arabic sentiments more than the identification of the polarity of English sentiments. In the three languages (Arabic, English, and Urdu), the researcher found out most of Arabic sentences began with a verb and the negative sentences were smaller than the positive sentences.

Al-Rubaiee et al. [41] studied the leading stock analysis software in Saudi Arabia Mubasher products and the relationship with Twitter in terms of opinion mining for trading strategies. The researchers built a model to derive feedback from Mubasher for sentiment analysis of MSA and DA tweets. The model was a combination of machine learning and natural language processing approaches to tag each tweet according to their sentiment polarity into: positive, negative or neutral. Moreover, AL-Rubaiee et al. [42] used different algorithms such as SVM, KNN and naïve Bayes during introducing Arabic text classification in stock trading.

The analysis showed results in SVM and KNN algorithms. However, the researchers have not used any type of correlations algorithms between the Saudi Arabian stock market indicator and Twitter, they have only used statistical measurements of differences to simply illustrate the relationship between Saudi Arabian TASI (TADAWUL ALL SHARE INDEX) and Twitter.

### 2.3    Twitter

Twitter API is the Application Programming Interface. The developers and researchers use API to complete a task by crawling or adjusting data. Relevant results to ad-hoc user queries provided by search API from a limited corpus or latest tweets. Moreover, API permit access to the nouns and verbs of Twitter such as writing tweets, following other users, reading timeline (Twitter [43]).

## 3    Corpus Collection

### 3.1    Data Pre-processing

We used Twitter API in order to retrieve Arabic financial related tweets. We generated a set of search queries to enhance the possibility of acquiring tweets that convey emotions, attitudes and opinions towards the specified entity. Crawling live stream tweets between the 27th of August 2015 and the 23rd of December 2015, the total volume of tweets collected was of about 277000.

It was designed to crawl all Arabic tweets regarding the Saudi Arabia stock market, TADAWUL, based on Arabic keywords such as (Table 1):

The tweets were stored in a local database, MongoDB, which contained objects of each tweet. During the crawling process, we added several filters to block possible spam, blacklisted spammers' accounts, removed duplicate tweets from the same ID and tweets containing long words. It was ensured that the source was safe, as there are many applications that generate automatic spam tweets. An initial black list of spammers IDs and key words of spamming tweets were gathered in previous test, as 5000 tweets were crawled based on the above and all spamming key words and account IDs were stored manually in black lists to reduce spamming in the crawling process Fig. 1.

**Table 1.** Keywords

| Arabic | Arabic Roman | Translation |
|--------|-------------|-------------|
| السوق | Assouq | the market |
| تاسي | TASI | TASI |
| تداول | TADAWUL | TADAWUL |
| الأسهم | Alashom | shares |
| السوق السعودي | Assouq Assaoodi | Saudi Market |
| سوق الأسهم | Sooq Alashom | stock market |
| سعر السوق | Se'er Assooq | Market Price |
| سعر الإغلاق | Se'er AleGhlaaq | closing price |
| ارتفاع | Ertifaa | Growth |
| هبوط | Hoboot | Fall |

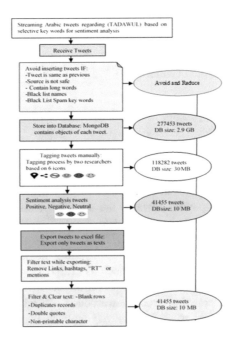

**Fig. 1.** Arabic tweets streaming API

The tweets contained public objects such as User ID, Screen Name, Name, created at, Text, Mention, Retweets, Following Count, Followers Count. We designed a platform for retrieving tweets from the database for simple statistical analysis and easier annotation process. In the next figure the graph illustrates the procedure of crawling, filtering, inserting, tagging and exporting Arabic tweets regarding TADAWUL.

**Table 2.** Tweets statistics

| Type of tweets | No. of tweets |
|---|---|
| Total tweets crawled | 277,453 |
| Total tweets tagged manually | 118,282 |
| Positive tweets | 5,449 |
| Negative tweets | 9,469 |
| Neutral tweets | 26,537 |
| Spam | 3,756 |
| Non-related meaning tweets | 58,840 |
| Non-Saudi Stock tweets | 14,231 |

**Fig. 2.** Annotation icons

Table 2 illustrates tweets statistics and the number of tweets analyzed manually the number of all tweets were inserted in the dataset, which shows that 277453 were collected. The researchers tagged around 118282 tweets into positive, negative and neutral tweets.

Figure 2 illustrates an Arabic tweet and the annotation icons used by the researchers, green refers to positive, red refers to negative, and orange refers to neutral. A red circle indicates it is spam, the 3-dots sign represents non-related tweets, not referring to the stock market. Location sign refers to stock market tweets related to another country not Saudi Arabia.

The tagging process of the collected tweets was a tedious process that took several months, because the tweets were of a very specific nature, the stock market, and from a specific country, Saudi Arabia. An Arab finance graduate researcher carried out the first round of the tagging process, because most of tweets contained finance related vocabulary and daily stock market information, afterwards a Saudi Arabian linguistics expert double checked the tweets tagged by the financial researcher.

As tagging process of the data was time consuming, the dataset was divided into two parts. The first part contained 118282 tweets and were analyzed for this paper; however, the second part is under tagging process, which will be released to the community in another research paper once the tagging and analysis are completed.

From the total tweets, 118283, as shown in Table 2, the usable ones for this study were only 41455 tweets, tagged as Positive, Negative and Neutral, 5449, 9469 and 26537 respectively. The unwanted tweets that were not considered for this study were 76827, divided as Spam 3756; None related meaning 58840 and Non KSA stock market 14231.

## 3.2    Data Analysis

Investors critically analyze past pricing history, which influences future investment decisions. social media and news items have a significant impact on stock market indices. In this paper, we applied machine learning and NLP principles to find the correlation between sentiments and trends in the stock market. A detailed process of machine learning analysis on tweets is described below.

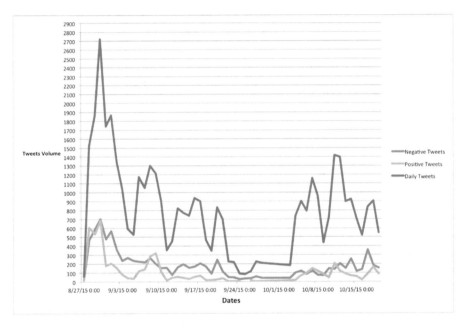

**Fig. 3.**  Volume of tweets daily from 27/08/2015 to 15/10/2015. Negative tweets are light blue, positive tweets is green, daily volume is purple color.

### 3.2.1    Data Preprocessing for Stock Market Trend Identification

The data set used for finding correlations between stock market trends and sentiments analysis, comprised of 41455 tweets posted between the 28th of August 2015 and the 24th of December 2015. Out of these 41455 tweets, 5499 tweets were categorized as positive tweets, 9469 as negative and 26537 as neutral. Figure 3 shows the graph of the total number of tweets posted daily along with the distribution of negative and positive tweets.

The Main attributes of the data set are user_ID, created_at, screen_name, followers_count, retweets_count, mentions, following count, text and status. The Table 3 provides definitions of these attributes.

Many fake accounts on social websites are used for spamming to influence the market with hidden purposes. For example, a twitter account following 180k accounts and having 190k followers is a fake account. We set up an empirical threshold of 5K following IDs. Hence the data set is filtered based on the user IDs who were following more than 5k accounts. User IDs with more than 5k accounts were placed in a separate data set categorized as Spam datasets and was excluded from the analysis phase. Filtered data set comprised of 38432 users who had the following_count of less than 5k.

A minority of members of our society possesses the quality to drive the trend changes in society and is known as the most influential users. Number of followers and number of retweets of their tweet and mentions are major variables in creating an influence. A tweet is categorized as the most influential tweet if it has many retweets and the market follows a positive or negative trend on that day or in the same week. Thus, comparing market performance with high influence tweets may result in identifying hidden patterns. In this paper, we have defined another empirical threshold regarding the most influential user whose tweet is retweeted more than 100 times. The filtered dataset was divided into two parts based on the most influential user. The first dataset consisted of all the accounts that had a retweet count greater than 100, and the status was positive. The second set comprised of accounts that had a retweet count greater than 100, and their status was labeled as negative.

**Table 3.** Main attributes of data set

| No. | Attribute name | Attribute description |
|-----|----------------|-----------------------|
| 1 | User_ID | Unique ID assigned by Twitter to each user |
| 2 | Created_at | Date and time on which tweet as posted |
| 3 | Screen_name | Account name displayed on twitter for each user |
| 4 | Followers_count | Number of other users following this account |
| 5 | Retweets_count | Total No of retweets posted for this tweet |
| 6 | Mentions | Total No of mentions of this tweet |
| 7 | Following_count | No of twitter accounts a user is following |
| 8 | Status | Polarity of the tweet text. It mainly consists of three emotions, negative, positive and neutral |
| 9 | Text | Arabic text 140 letters |

The Saudi Arabian Stock Market Exchange, referred as TADAWUL, is used in this paper for finding trends. Closing prices for TADAWUL All Share Index (TASI) were collected through TADAWUL website www.tadawul.com.sa. The closing prices for TASI from the 2nd of August 2015 to the 29th of October 2015 is used to predict relationships between the most the influential users, tweets and the stock market prices. Figure 4 shows the distribution of TASI performance.

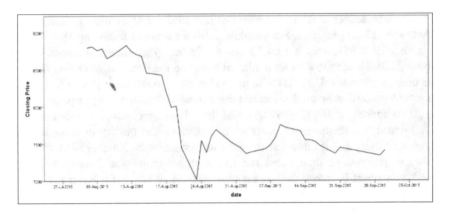

**Fig. 4.** TASI performance from the end of June 2015 to the beginning of Oct 2015 with closing price between 7000 points to 9000 points.

### 3.2.2   Methodology for Comparing User Influence with Stock Market Prices

Before the application of machine learning algorithms, we plotted graphs for the most influential users' retweet counts and TASI performance on the same dates, to manually observe the correlation. Figure 5 illustrates the rise and fall of retweet counts on certain dates based on the status type. It is observed from the graph that the tweets that were assigned a negative status increased the number of retweet counts from the 10th to the 14th of September 2015. Figure 6 shows a decrease in the Closing prices from the 10th to the 14th of September 2015. Hence there is a direct correlation between the variable retweet count and the closing price.

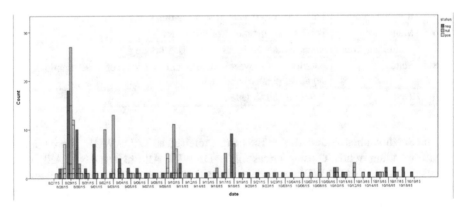

**Fig. 5.** Retweet counts based on negative, positive and neutral status, blue is negative retweets, yellow is positive retweets, green is neutral.

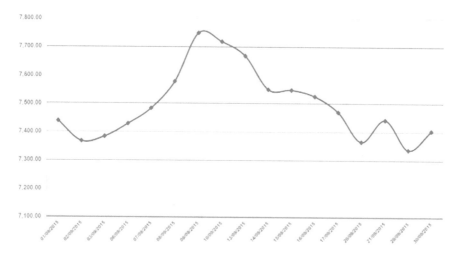

**Fig. 6.** TASI closing price distribution for the month of September 2015

Correlation analysis is a bivariate analysis that measures the strengths of association between two variables. Correlation is categorized into three categories by considering the effect of the rise in one variable on another variable (Bobko [44]).

1. Positive Correlation: another variable is also rising. For a significant correlation between two variables, its value should be greater than 0.5 and less than 1.
2. Negative Correlation: another variable is decreasing. The value of coefficient is less than 0.5.
3. No Correlation: Neither increase nor decrease in other variable.

Following three types of correlation coefficients were used to find associations between two variables.

1. Pearson's correlation coefficient is a statistical measure for calculating the dependence between two linearly related variables. It is denoted by $\Gamma$ and the following equation shows the range for $\Gamma$.

$$-1 \le \Gamma \le 1$$

The formula for Pearson's correlation coefficient is given by equation below.

$$\Gamma = \frac{\sum XY - \frac{\sum X \sum Y}{n}}{\sqrt{\left(\left(\sum X^2 - \frac{\sum X^2}{n}\right)\left(\sum Y - \frac{\sum Y^2}{n}\right)\right)}} \qquad (1)$$

Where
$\Gamma$ = *Pearson Correlation Coefficient*
*n* = *Number of values in each data set*
*X* = *Variable* 1
*Y* = *Variable* 2
For Pearson correlation coefficient, both variables should have normal distributions (Peter et al. [45]).

2. Kendall rank correlation is a non-parametric test used for measuring the strength of association between two variables. The following equation describes the formula for calculating the Kendall rank coefficient (Kendall et al. [46]).

$$\tau = \frac{n_c - n_d}{\frac{1}{2}n(n-1)} \tag{2}$$

Where
$n_c$ = *Number of Concordant*
$n_d$ = *Number of discordant*

3. Spearman rank correlation is a non-parametric test used for finding relationships between two variables. This coefficient does not assume the distribution of data and is appropriate for ordinal variables. The formula for calculating this coefficient is given below (Peter et al. [45]).

$$\rho = 1 - \frac{6\sum d_i^2}{n(n-1)} \tag{3}$$

Where
$\rho$ = *Spearman Rank Coefficient*
*n* = *Number of Values in each Data set*
$d_i$ = *The difference between rank of corresponding values of X and Y.*

## 4   Results

Pearson's correlation coefficient is applied the data set comprised of negative status and closing prices, yield the result shown in Table 4.

**Table 4.** Pearson's correlation

| Pearson's coefficient | Retweet_count | TASI closing price |
|---|---|---|
| Retweet_count | 1.000 | .022 |
| Sig (1Tailed) | | .463 |
| N | 20 | 20 |
| TASI Closing Price | .022 | 1.000 |
| Sig (1Tailed) N | .463 | |
| | 20 | 20 |

Results of Spearman's and Kendall's correlation coefficients are shown in Tables 5 and 6 below.

**Table 5.** Spearman's correlation

| Spearman's coefficient | Retweet_count | TASI closing price |
|---|---|---|
| Retweet_count | 1.000 | .197 |
| Sig (2Tailed) | | .404 |
| N | 20 | 20 |
| TASI closing price | .197 | 1.000 |
| Sig (2Tailed) N | .404 | |
| | 20 | 20 |

**Table 6.** Kendall's correlation

| Kendall's coefficient | Retweet_count | TASI closing price |
|---|---|---|
| Retweet_count Sig (1Tailed) | 1.000 | .142 |
| N | | .218 |
| | 20 | 20 |
| TASI closing price Sig (1Tailed) | .142 | 1.000 |
| N | .218 | |
| | 20 | 20 |

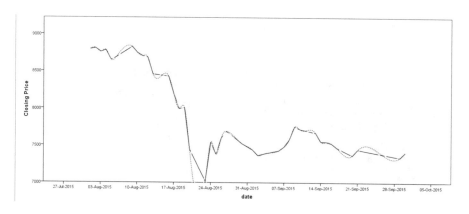

**Fig. 7.** Interpolation for TASI performance from 27th June to 5th October.

The value of all three correlation coefficients is less than 0.5 because there are certain dates on which TASI closing prices are missing. Linear Interpolation is used for replacing the missing values with the interpolated values.

Linear Interpolation is the process of fitting curve using linear polynomial to construct new data points within the discrete data set (Hazewinkel et al. [47]). Graph Fig. 7 shows the interpolation for the month of Aug 2015 and Sept 2015.

Furthermore, there are certain dates on which increase in retweets_count with negative status, does not result in decrease in closing price for TASI performance. The random number of retweet_count for different tweets on a certain date results in decrease in value of Pearson's coefficient. Hence retweets_count with negative status on same date are grouped and summation is used for creating new variable total_retweet_count per day. Figure 8 shows the result for grouping retweets_count on same day with negative status.

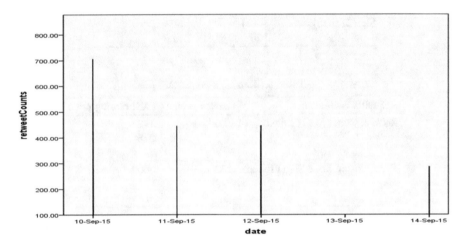

**Fig. 8.** Total negative retweets_count from 10th September to 14th September

**Table 7.** Pearson's correlation for total_retweet_count and TASI closing price.

| Pearson's coefficient | Total_retweets_count | TASI closing price |
|---|---|---|
| Total_retweets_count | 1.000 | .465 |
| Sig (2Tailed) | | .430 |
| N | 5 | 5 |
| TASI Closing Price | .465 | 1.000 |
| Sig (2Tailed) | .430 | |
| N | 5 | 5 |

Pearson's correlation coefficient is calculated for total_retweets_count and closing price. Results are shown in Table 7. A significant increase in the value of Pearson's correlation coefficient and other coefficients is observed still it is less than 0.5 due to small dataset and missing value on 13th Sept 2015. To further improve the results, missing value on 13th Sept 2015 is calculated using linear interpolation and Pearson's correlation coefficient is applied. A significant improvement in the value of Pearson's, Spearman's and Kendall's correlation coefficient is observed as shown below (Tables 8, 9 and 10).

**Table 8.** Pearson's correlation after linear interpolation.

| Pearson's coefficient | Total_retweets_count | TASI closing price |
|---|---|---|
| Total_retweets_count | 1.000 | .646 |
| Sig (2Tailed) | | .239 |
| N | 5 | 5 |
| TASIClosing Price | .646 | 1.000 |
| Sig (2Tailed) N | .239 | |
| | 5 | 5 |

**Table 9.** Spearman's correlation after linear interpolation

| Spearman's coefficient | Total_retweets_count | TASI closing price |
|---|---|---|
| Total_retweets_count | 1.000 | .667 |
| Sig (2Tailed) | | .219 |
| N | 5 | 5 |
| TASI Closing Price | .667 | 1.000 |
| Sig (2Tailed) N | .219 | |
| | 5 | 5 |

**Table 10.** Kendall's correlation after linear interpolation

| Kendall's coefficient | Total_retweets_count | TASI closing price |
|---|---|---|
| Total_retweets_count | 1.000 | .527 |
| Sig (2Tailed) | | .207 |
| N | 5 | 5 |
| TASI Closing Price | .527 | 1.000 |
| Sig (2Tailed) | .207 | |
| N | 5 | 5 |

Furthermore, the variable mentions play an important role in identifying those twitter accounts, whose tweet contributed towards market trends. The graph Fig. 9 shows the count for mentions of twitter accounts. Four major twitter accounts are identified based on mentions count greater than 100. Identity of these accounts is kept hidden due to privacy concerns. A histogram analysis of these users based on number of mentions on certain dates identifies that tweets posted by these users have a major impact on stock market prices (Figs. 10 and 11).

The tweets posted by user A and user B have major influence on stock market shares as these users are mentioned frequently on the same dates when the stock market prices following a decreasing trend (Figs. 12 and 13).

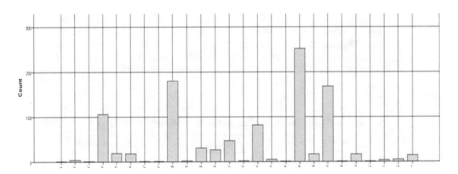

**Fig. 9.** Histogram analysis for Twitter users based on their tweet's mentions and count.

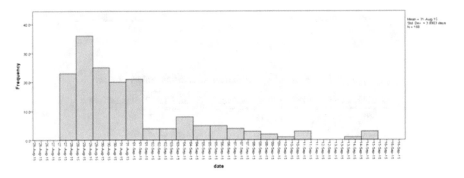

**Fig. 10.** Count of mentions for User A on different dates.

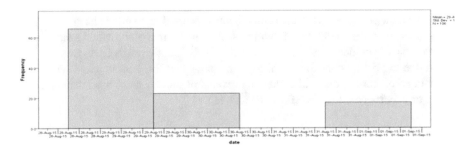

**Fig. 11.** Count of mentions for User B on different dates.

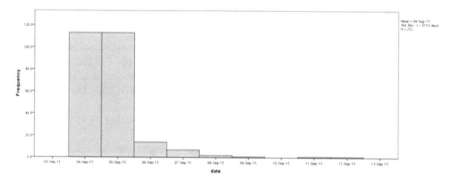

**Fig. 12.** Count of mentions for User C on different dates

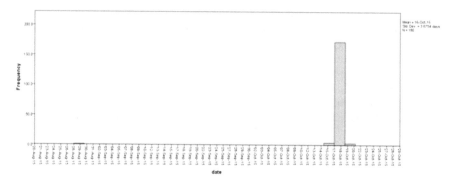

**Fig. 13.** Count of mentions for User D on different dates

## 5   Conclusion

This paper analyzed the influence of Twitter users on TASI performance. We found that there is direct correlation between status, retweets_count and closing prices of the stock market. The most influential users, based on their retweets_count, can hold significant influence over the rise and fall of TASI performance. Furthermore, it found that the variable mentions play an important role in identifying twitter accounts that drive stock market trends.

This means that the most influential users are predictable, and in the future such users can have a significant impact on the stock market trends.

**Acknowledgments.** This research is supported by The National Natural Science Foundation of China with Grant No: 61272277.

We would also like to thank RIC at PSU for their support.

# References

1. Maia, M., Almeida, J., Almeida, V.: Identifying user behavior in online social networks. In: Proceedings of the 1st Workshop on Social Network Systems, pp. 1–6 (2008)
2. Mislove, A., Marcon, M., Gummadi, K.P., Druschel, P., Bhattacharjee, B.: Measurement and analysis of online social networks. In: Proceedings of the 7th ACM SIGCOMM Conference on Internet Measurement, pp. 29–42 (2007)
3. Bao, Y., Quan, C., Wang, L., Ren, F.: The Role of Pre-processing in Twitter Sentiment Analysis, pp. 615–624. Springer, Cham, (2014)
4. Chen, Y.-J., Chen, Y.-M., Lu, C.L.: Enhancement of stock market forecasting using an improved fundamental analysis-based approach. Springer, Heidelberg (2016)
5. Oliveira, N., Cortez, P., Areal, N.: On the Predictability of Stock Market Behavior Using StockTwits Sentiment and Posting Volume. Springer, Heidelberg (2013)
6. Ho, K.-Y., (Walter) Wang, W.: Predicting Stock Price Movements with News Sentiment: An Artificial Neural Network Approach. Springer International Publishing Switzerland (2016)
7. Qasem, M., Thulasiram, R., Thulasiram, P.: Twitter sentiment classification using machine learning techniques for stock markets. In: International Conference on Advances in Computing, Communications and Informatics (ICACCI) (2015)
8. Cha, M., Benevenuto, F., Haddadi, H., Gummadi, P.K.: The world of connections and information flow in Twitter. IEEE Trans. Syst. Man Cybern. Part A **42**(4), 991–998 (2012)
9. Abdul-Mageed, M., Diab, M.: AWATIF: a multi-genre corpus for modern standard Arabic subjectivity and sentiment analysis. In: Proceedings of the Eight International Conference on Language Resources and Evaluation (LREC'12), Istanbul, Turkey. European Language Resources Association (ELRA) (2012)
10. Al-Sabbagh, R., Girju, R.: YADAC: yet another dialectal Arabic corpus. In: Proceedings of the Eight International Conference on Language Resources and Evaluation (LREC'12), Istanbul, Turkey. European Language Resources Association (ELRA) (2012)
11. Refaee, E., Rieser, V.: Subjectivity and sentiment analysis of arabic twitter feeds with limited resources. In 9th International Conference on Language Resources and Evaluation (LREC'14), (2014)
12. Pang, B., Lee, L., Vaithyanathan, S.: Thumbs up? Sentiment classification using machine learning techniques. In: Proceedings of the ACL 2002 Conference on Empirical Methods in Natural Language Processing, vol. 10, pp. 79–86. Association for Computational Linguistics (2002)
13. Aciar, S., Zhang, D., Simoff, S., et al.: Informed recommender: basing recommendations on consumer product reviews. IEEE Intell. Syst. **22**(3), 39–47 (2007)
14. Kang, H., Yoo, S.J., Han, D.: Senti-lexicon and improved Naïve Bayes algorithms for sentiment analysis of restaurant reviews. Expert Syst. Appl. **39**(5), 6000–6010 (2012)
15. Atsalakis, G.S., Valavanis, K.P.: Surveying stock market forecasting techniques - part II: soft computing methods. Expert Syst. Appl. **36**(3), 5932–5941 (2009)
16. Silva, E., Castilho, D., Pereira, A., Brando, H.: A neural network-based approach to support the market making strategies in high-frequency trading. In: Proceedings of the International Joint Conference on Neural Networks, pp. 845–852 (2014)
17. Girijia, V.A., Manohara, P.M.M., Radhika, M.P., Aparna, K.: Stock Market Prediction: A Big Data Approach. IEEE (2015)
18. Fama, E.F., et al.: The adjustment of stock prices to new information. Int. Econ. Rev. **10**(1), 1–21 (1969)
19. Korayem, M., Crandall, D., Abdul-Mageed, M.: Subjectivity and Sentiment Analysis of Arabic: A Survey, pp. 128–139. Springe, Heidelberg (2012)

20. Rushdi-Saleh, M., Martín-Valdivia, M.T., Ureña-López, L.A., Perea-Ortega, J.M.: OCA: opinion corpus for Arabic. J. Am. Soc. Inform. Sci. Technol. **62**(10), 2045–2054 (2011)
21. Elarnaoty, M., AbdelRahman, S., Fahmy, A.: A Machine Learning Approach For Opinion Holder Extraction Arabic Language. CoRR, abs/1206.1011 (2012)
22. Elhawary, M., Elfeky, M.: Mining Arabic business reviews. In: Proceedings of International Conference on Data Mining Workshops (ICDMW), pp. 1108–1113. IEEE (2010)
23. El-Halees, A.: Arabic opinion mining using combined classification approach. In: Proceedings of the International Arab Conference on Information Technology, ACIT (2011)
24. Ibrahim, H.S., Abdou, S.M., Gheith, M.: MIKA: a tagged corpus for modern standard Arabic and colloquial sentiment analysis. In: The 2nd IEEE International Conference on Recent Trends in Information Systems (ReTIS) (2015)
25. Refaee, E., Rieser, V.: An Arabic Twitter Corpus for Subjectivity and Sentiment Analysis (2015)
26. Nabil, M., Aly, M., Atiya, A.F.: ASTD: Arabic sentiment tweets dataset. In: Proceedings of the 2015 Conference on Empirical Methods in Natural Language Processing, Lisbon, Portugal, 17–21 September 2015, pp. 2515–2519 (2015)
27. Duwairi, R.M., Marji, R., Sha'ban, N., Rushaidat, S.: Sentiment analysis in Arabic tweets. In: The 5th International Conference on Information and Communication Systems (ICICS) (2014)
28. Huang, Y., Zhou, S., Huang, K., Guan, J.: Boosting Financial Trend Prediction with Twitter Mood Based on Selective Hidden Markov Models, pp. 435–451. Springer, Cham (2015)
29. Idvall, P., Jonsson, C.: Algorithmic trading: hidden markov models on foreign exchange data. Master's thesis, Sodertorn University (2008)
30. Li, X., Wang, C., Dong, J., Wang, F., Deng, X., Zhu, S.: Improving stock market prediction by integrating both market news and stock prices. In: Hameurlain, A., Liddle, S.W., Schewe, K.-D., Zhou, X. (eds.) DEXA 2011, Part II. LNCS, vol. 6861, pp. 279–293. Springer, Heidelberg (2011)
31. Pidan, D., El-Yaniv, R.: Selective prediction of financial trends with hidden markov models. In: Advances in Neural Information Processing Systems, pp. 855–863 (2011)
32. Zhang, Y.: Prediction of financial time series with Hidden Markov Models. Master's thesis, Simon Fraser University (2004)
33. Bollen, J., Mao, H., Zeng, X.: Twitter mood predicts the stock market. J. Comput. Sci. **2**(1), 1–8 (2011)
34. Mittal, A., Goel, A.: Stock prediction using twitter sentiment analysis. Technical report, Stanford University
35. Si, J., Mukherjee, A., Liu, B., Li, Q., Li, H., Deng, X.: Exploiting topic-based twitter sentiment for stock prediction. In: Proceedings of the 51st Annual Meeting of the Association for Computational Linguistics (vol. 2, Short Papers), pp. 24–29 (2013)
36. Sprenger, T.O., Tumasjan, A., Sandner, P.G., Welpe, I.M.: Tweets and trades: the information content of stock microblogs. European Financial Management (2013)
37. Wu, D., Ke, Y., Yu, J.X., Yu, P.S., Chen, L.: Detecting leaders from correlated time series. In: Kitagawa, H., Ishikawa, Y., Li, Q., Watanabe, C. (eds.) DASFAA 2010. LNCS, vol. 5981, pp. 352–367. Springer, Heidelberg (2010)
38. Yang, B., Guo, C., Jensen, C.S.: Travel cost inference from sparse, spatio temporally correlated time series using markov models. Proc. VLDB Endow. **6**(9), 769–780 (2013)
39. Ahmad, K., Cheng, D., Almas, Y.: Multilingual sentiment analysis of financial news streams. In: Proceedings of the 1st International Conference on Grid in Finance, Palermo, pp. 1–8 (2006)

40. Almas, Y., Ahmad, K.: A note on extracting 'sentiments' in financial news in English, Arabic & Urdu. In: The 2nd Workshop on Computational Approaches to Arabic Script-Based Languages, Linguistic Soc America 2007 linguistic Institute, Stanford University, Stanford, California, Linguistic Society of America, pp. I–12 (2007)

41. AL-Rubaiee, H., Qiu, R., Li, D.: Identifying Mubasher Software Products through Sentiment Analysis of Arabic Tweets, Crown (2016). 978-1-4673-8743-9/16/

42. AL-Rubaiee, H., Qiu, R., Li, D.: Analysis of the relationship between Saudi twitter posts and the Saudi stock market. In: IEEE Seventh International Conference on Intelligent Computing and Information Systems, ICICIS 2015 (2015)

43. Twitter. The Search API (2016). https://dev.twitter.com/rest/public/. Accessed 15 June 2016

44. Bobko, P.: Correlation and Regression: Applications for Industrial Organizational Psychology and Management, 2nd edn. Sage Publications, Thousan Oaks (2001)

45. Chen, P.Y., Popovich, P.: Correlation: Parametric and Non-parametric Measures. Sage Publications, Thousand Oaks (2002)

46. Kendall, M.G., Gibbons, J.D.: Rank Correlation Methods, 5th edn. Edward Arnold, London (1990)

47. Hazewinkel, M. (ed.): Linear Interpolation. Encyclopedia of Mathematics, Springer (2001). ISBN 978-1-55608-010-4

# Comparative Analysis of Danger Theory Variants in Measuring Risk Level for Text Spam Messages

Kamahazira Zainal[1(✉)] 🆔, Mohd Zalisham Jali[1] 🆔,
and Abu Bakar Hasan[2] 🆔

[1] Faculty of Science and Technology, Universiti Sains Islam Malaysia (USIM),
71800 Nilai, Negeri Sembilan, Malaysia
arizah78@yahoo.com, zalisham@usim.edu.my
[2] Faculty of Engineering and Built Environment,
Universiti Sains Islam Malaysia (USIM),
71800 Nilai, Negeri Sembilan, Malaysia
abakarh@usim.edu.my

**Abstract.** The issue of spam has been uprising since decades ago. Impact loss from various aspects has attacked the daily life most of us. Many approaches such as policy and guidelines establishment, rules and regulations enforcement, and even anti-spam tools installation appeared to be not enough to restrain the problem. To make things even worse, the spam's recipients still easily get enticed and lured with the spam content. Hence, an advanced medium that acts as an implicit decision maker is desperately required to assist users to obstruct their eagerness responding against spam. The simulation of spam risk assessment in this paper is purposely to give some insights of how users can identify the imminent danger of received text spam. It is demonstrated by predicting the potential hazard with three different levels of risk (high, medium and low), according to its possible impact loss. A series of simulation has been conducted to visualize this concept using Danger Theory variants of Artificial Immune Systems (AIS), namely Dendritic Cell Algorithm (DCA) and Deterministic Dendritic Cell Algorithm (dDCA). The corpus of messages from UCI Machine Learning Repository has been deployed to illustrate the analysis. The outcome of these simulations verified that dDCA has consistently outperformed DCA in precisely assessing the risk level with severity concentration value for text spam messages. The findings of this work has demonstrated the feasibility of immune theory in risk measurement that eventually assisting users in their decision making.

**Keywords:** Dendritic Cell Algorithm (DCA)
Deterministic Dendritic Cell Algorithm (dDCA) · Risk classification
Signals processing · Text mining · Text spam risk assessment

## 1 Introduction

Spam attack has become very well known in email platform. This issue has risen for quite some time and persists along with its negativity. Even until now, there is no available foolproof system yet that completely effective to prevent spam. Exacerbating

© Springer International Publishing AG, part of Springer Nature 2018
M. Alenezi and B. Qureshi (Eds.): *5th International Symposium
on Data Mining Applications*, pp. 133–152, 2018.
https://doi.org/10.1007/978-3-319-78753-4_11

the situation, there is a rapid growth of spam technology to evade the anti-spam technology, surprisingly within a few months [1]. The booming of mobile technology has worsened the impact of spam which has expanded the territory of spam from email to SMS, social media, and website platform. Even more, its impact has evolved from as minimal of losing productivity, inefficiency infrastructure usage, and financial gain to criminal activities [2, 3].

Currently, numerous efforts have been established to overcome this issue. These include the development of policy and standards, the establishment of information commissioner to report spam at the national and global level, installing anti-spam software and even incorporate education program on security awareness for users, which has been discussed comprehensively in [4]. The spam problem could not be solved successfully unless every user practices good ethics while using the technical facilities of Internet chains, which in particular that include smartphones and social media. Although various methods have been introduced in the form of technology and legislation, the spam problem is still increasing. Technology and legislation itself could not eliminate the spam because it needs the cooperation of the user of which, anybody potentially becomes the victim. Users' behavior on how they react against spam somehow also influences the success rate of spam intention.

Motivated by this unending conflict, the availability of a tool in helping users decide on how to react to spam is critically needed. Hence, there is a study required to develop this tool in assisting users to understand and foreseen the inherent danger hidden in a spam message. This paper objectively covers some parts of the tool's design which is to verify the capability of Danger Theory (DT) and its variants to assess the risk of text spam. The verification process is conducted via a series of simulation. This instrument is impersonating the idea of how human biological immune cells are functioning and operating naturally to measure the severity level of a risk for text spam messages. Keen to probe the possibility that this idea of DCs can potentially assist to measure the risk concentration, a series of experiments have been conducted to validate this theory in the domain of spam risk assessment. Although many types of research have been carried out for the detection of spam, no publications have been found so far for assessing the risk level of this threat, specifically by applying AIS. Besides differentiating the risk into a few distinguish level, the measured risk level is also referring to the concentration of possible impact loss measured in numerical value.

In reality, most users are not aware of the implications of spam problem because the consequences are not visible. This paper purposely will elaborate how the proposed model solution would assist the user to identify the danger that could emerge from a spam message. It is intended to produce a tool that able to help the user to distinguish spam that is potentially severe by measuring its risky content. By knowing and realize the impact that could emerge caused by this spam, the user is expected to be more careful and not to rely on the deceitful content. The proposed model defined the context of spam in 3 different risk levels according to the measured severity concentration.

For an appropriate content arrangement, this paper is structured as follows. Related works that become the motivation of this study is covered in Sect. 2. Then, the idea of Danger Theory that has become the fundamental theory and as the methodology for the proposed solution is elaborated in Sect. 3. In Sect. 4, the setup for the series of experiment and its objective is explained. Results and analysis of all experiments are presented in Sect. 5. The findings also discussed in this particular section. Finally, the areas for potential future works are presented in Sect. 6.

## 2    Related Works

### 2.1    Spam and Human Reaction Against It

Spam problem has been with us since the year of the 1990s and yet there is no available method that capable of combating and diminish this threat in total. Although many anti-spam solutions have been developed via various research and studies, it is only cover the filtering or the detection of spam. At this time of writing, there is no publication found so far for managing spam as a risk.

Spam is observed as one of the mobile threat based on its malevolent impacts, as discussed by Theoharidou *et al.,* [2] and Yeboah-Boateng *et al.,* [5]. Subsequently, a study proposed for a solution has been shared by Zainal *et al.,* in [6]. This study discussed how this mobile threat can be managed as suggested in many developed and practiced standards of risk management. The solution proposed three main processes that involve in managing spam which include spam classification, spam clustering, and the final process is determination level of spam's severity that implicitly notifies users of possible hidden danger lies in a spam message.

Humans are typically the weakest link [5] in any security system. In spite of the fact that technology and technical sophistication in spam avoidance, cybercriminals often try to exploit human weaknesses as a way of spreading their spam messages. Users are not only susceptible due to a lack of awareness, but sometimes the lure of the offer can entice people into clicking on a link that should just be ignored.

Therefore, an essential method is required as one of the possible ways to educate users about human dimension and reaction in information security. The cause and effect affiliation must always be well aware of users in electronic communication, to protect themselves from online and mobile threats. Users also must be alerted that there are available of established legislation and law enforcement initiatives to make sure they are safe in this digital world.

Hence, the principal objective of this work is to develop a solution in assisting users to be well aware of the danger hidden in a spam message. Once they realized with the malicious intent of spam, this will help users decide on how to react to spam.

### 2.2    Biological-Inspired Application in Computational Intelligence

Inspired by Body Immune System (BIS), this study proposed a solution based on biological properties. A theory that is manipulating the way human body defense from unknown and danger particles has been initiated about a couple of decades ago, which also has been implemented successfully in a real-world problem. This theory that is known as the Artificial Immune System (AIS), imitate human Body Immune System (BIS) to protect the body from any detected danger that could bring severe harm, has been clearly proven in practice of computer security. Some domain of computer security that has practically employed AIS as the potential solution includes intrusion detection, spam classification, risk assessment, robotics, and machine learning. Reviews and conceptual survey of this implementation can be found in [7–11].

Read *et al.*, [12] clarified that the mechanisms of innate immunity are generic defense systems that are non-specific to a particular pathogen, but act against general types of pathogen. Comparatively, for the mechanisms of adaptive immunity facilitate the immune system to adjust earlier unseen pathogens based upon exposure to them. AIS algorithms masquerade the behavior and functions of biological properties of immunological cells, to be specific B-cells, T-cells and Dendritic Cells (DCs).

This study is imitating the behavior of DCs which are the fundamental possession of Danger Theory. They are truly an intrusion or anomaly detection agent in the human body and acting as messengers between the innate and the adaptive immune systems.

## 3    Methodology

### 3.1    Danger Theory

The Danger Model was initially proposed by Polly Matzinger, an immunologist, who suggested a novel explanation of how the immune system works. She proposed Danger Model [13], which suggests that the immune system is more concerned with the damage than cell foreignness. This idea then adapted to computational intelligence as Danger Theory that becomes known as the second generation of AIS.

In 2007, Greensmith applied this Danger Theory to a series of an experiment for intrusion detection that is implemented and elaborated in her dissertation [14]. The intrusion detection and anomaly detection was the original works in the adaptation of Dendritic Cell Algorithm or DCA as a biologically inspired idea in computer security domain.

DCA become the prominent algorithm which is a signal processing system that is inspired by the behavior of dendritic cells (DCs) [14]. DCs are antigen presenting cells (APCs) which play a critical role in the regulation of the adaptive immune system. DCs are unique APCs that are capable of capturing, processing and presenting antigens on the cell surface along with appropriate co-stimulation (CSM) molecules.

This paper is focusing only on the initial version of DCA and dDCA. dDCA or also known as Deterministic Dendritic Cell Algorithm is a less complex version than another could potentially minimize the computational cost since a large amount of randomness has been removed. DCA and dDCA are much simpler than other variants, and dDCA has reduced numbers of parameters compared to DCA. Authors Greensmith *et al.*, [15] and Brownlee [16] identified that complexity had been reduced in dDCA to make it more feasible and amenable to analysis. Even though numerous parameters have been removed, there is no reduction in algorithm performance as clarified in [17].

The essence of both DCA and dDCA is to correlate disparate data streams in the form of antigens and signals and label each group of identical antigens as anomalous or normal. These DCs are possessing in the body tissues and gather antigen and other signals (danger) that depicting the current condition of the tissues. This representation of the current state portrayed if the antigen has been gathered in a safe or dangerous context, and causes DCs to change into a semi-mature (smDC) or mature (mDC) state.

The mission of the DCs is to distinguish antigens as being either benign (harmless) or malignant (harmful) in nature [18].

Aickelin *et al.,* [19] emphasized that within the biological systems, Pathogen Associated Molecular Patterns (PAMPs) are molecules released exclusively by pathogens. Then, danger signals are released from tissue cells following unplanned necrotic cell death, while safe signals are released from normally dying cells as an indicator of healthy tissue. It concurs that PAMPs indicate a higher level of hazard compared to danger signal. Inflammation is classed as the molecules of an inflammatory response to tissue injury. These signals are classified as input signals that are further processed to produce an output signal that describes the surrounding condition.

The signals that migrated to the lymph node are divided into two types of signals as regards to the degree or concentration of danger detected. Apoptotic alerts or semi-mature brings the safe signal, while necrotic alerts bring the mature signal. Semi-mature indicate a 'safe' context and mature indicate a 'dangerous' context. These informative signals are a reflection of the current state of the surrounding [18].

The collected data of antigen by DCs are measured using the Mature Context Antigen Value (MCAV) which this is the mean value of context per antigen type. In other words, this MCAV determines the intensity or degree of the detected danger. This can be measured by the following equation:

$$MCAV\ (antigen\_type) = \frac{mature\_count}{antigen\_count} \tag{1}$$

As clarified by Greensmith *et al.,* [20], the closer this value to 1 means the greater the probability that the antigen is anomalous. This MCAV value is used to assess the degree of the anomaly of an antigen.

Besides, a detailed elaboration about migration threshold and anomaly threshold has been discussed in [21]. The DCA introduces migration threshold to determine the lifespan of the DC. Whenever cumulative CSM exceeds the migration threshold, the DC ceases to sample signals and antigens. At this point, the other two cumulative smDC and mDC are assessed. While for anomaly threshold, the chosen value reflects the distribution of normal and anomalous items presented within the original data set. Antigens with MCAV which exceeds this threshold are classified as anomalous and vice versa.

The calculation scheme for the context assessment to determine its anomalous level is a bit different between DCA and dDCA [15]. In the dDCA, the anomaly metric, $K_\alpha$ is implemented and the magnitude of k value is used. This generates real-valued anomaly scores and may assist in the polarization of normal and anomalous processes. The process of calculating this anomaly score is shown in Eq. (2), where $k_m$ is the k value for $DC_m$, $\alpha_m$ is the number of antigens presented of type $\alpha$ by $DC_m$.

$$K_\alpha = \frac{\sum_m k_m}{\sum_m \alpha_m} \tag{2}$$

Other than that, there is a minor differentiation in translating the value of anomalous level in dDCA [22]. The outcome of $K_\alpha$ in dDCA is tagged as anomalous when it is returned as a positive value, $K_\alpha > 0$ and tagged as normal when the returned value is negative, $K_\alpha < 0$.

In spite of that, DCA shows a relationship of cause and effect that is believed exists between signals and antigens, where signals are the explicit effects that potentially result from the implicit cause of antigens. The signal profile is a measure of processed signal instances, whereas the antigen profile is a measure of sampled antigen instances.

### 3.2    Algorithms and Integration with Text Mining

In this work, the design principles for constructing the proposed risk assessment system is according to identified biological behavior and property of interest of DCs. An initial algorithm for this work has been proposed in [6] which a general concept of designing and developing a risk assessment method for text spam message is presented. The concept is articulated on how the approach from risk management and Danger Theory can be integrated. Then, an additional approach [23] from text mining is proposed to be integrated with this proposed system.

The initial algorithm as proposed in [6] is further designated for DCA and dDCA, as depicted in Figs. 1 and 2. These enhanced pseudo-codes are clarified thoroughly in [24].

Line 8–14 for both classifiers are referring to the integration with text mining approach, which covered the pre-processing phase and the calculation of term weighting that stipulated as an input signal. While for line 15 is to illustrate the integration of risk scale in order to distinguish three different levels of input and output signals. These two figures have shown a clear differentiation of spam risk measurement between DCA and dDCA in line 17 onwards.

As identified in many papers, pre-processing reduced high dimensional data that produce noise, hence this process commonly assists classifier to perform more effective and require low computational time. The process is further tested and its effect is identified and elaborated in [23, 25]. In addition to that, term weighting scheme is applied to calculate the weight of every term in the context using data mining tool, RapidMiner, version 5.3. Term Frequency (TF), Information Gain Ratio (IG Ratio) and CHI Square ($CHI^2$) have been utilized as the pre-feature selection methods. The value of every term derived from this method subsequently deployed as input signals. In most existing research on spam identification, these three schemes have been validated as the most outstanding method. Due to this reason, the same methods are deployed in this study to inspect its potentiality in collaboration with the DCA and dDCA classifier to measure the risk concentration. Some existing works that deploy these three schemes in spam identification can be found in [25–29].

### 3.3    Dataset

For the initial population, a corpus from UCI Machine Learning Repository [30] has been deployed. With reference to [23], this public-shared corpus is the most used by many researchers around the world for various associated tasks of research, besides of utilizing synthetic and self-collect data set. To allow any improvement of the methodology employed and able to compare for identifying the reliable solution, this study also uses the same public-shared corpus.

```
1    input      : message dataset (spam and ham) and pre-categorized signals (weights)
2    output     : spam message with risk concentration value
3    parameter : anomaly threshold
4
5    initialize assessment;
6    while dataset of messages available do
7            get messages
8            tokenize message
9            calculate weights for signal generation
10           store assigned weights for tokenized words
11   endwhile
12   for every new spam message do
13           tokenize message
14           choose weighting scheme
15           set value for risk scale with 3 different levels and weight to transform input signal to
             output signal
16           identify for the maximum, minimum and average value of weights
17                   calculate OUTPUT SIGNALS, O[CSM, smDC, mDC] for 3 cycles; maximum, minimum
                     and average signals value
18                   if O[mDC] > O[smDC] then,
19                           spam message is assigned as 1 to indicate the malicious message; and
20                           MCAV result should calculate as High or Medium
21                   else
22                           spam message is assigned as 0 to indicate the benign message; and
23                           MCAV result should calculate as Low
24           identify words with high, medium and low signals value
25                   calculate MCAV
26           identify anomaly threshold, tm
27                   if MCAV > anomaly threshold, tm then,
28                           spam message is anomalous
29                   else
30                           spam message is normal
31                   endif
32                   print spam message with the MCAV concentration value and its associated risk level
33   end
```

**Fig. 1.** The enhanced pseudo-code of DCA for measuring the risk level of text spam messages.

## 4  Experimental Setup

### 4.1  Aims

The concept paper for this study of spam risk measurement has been proposed in [6] which articulated the link of biological properties in Danger Theory with the message spam environment. The conceptual framework proposed in paper [6] is then tested with a series of simulation conducted in [25]. The findings from this initial simulation is later applied in the designation and development of the prototype, as elaborated in [24].

The work presented in [25] is then extended with the testing executed in this paper. The objective of this initial work [25] is to identify the best term weighting scheme, risk scale and signal weight matrix that optimized producing high true positive value in classifying the risk level of text spam messages using DCA classifier only. Other than identifying the best term weighting scheme, risk scale and signal weight matrix, the count for tokens for anomaly measurement also has been tested. As described in Eq. (1), the anomaly measurement is based on MCAV calculation. For this purpose, the mature content of the antigen must be identified for the calculation. Two conditions

```
1     input      : message dataset (spam and ham) and pre-categorized signals (weights)
2     output     : spam message with risk concentration value
3     parameter  : anomaly threshold
4
5     initialize assessment;
6     while dataset of messages available do
7             get messages
8             tokenize message
9             calculate weights for signal generation
10            store assigned weights for counted words
11    endwhile
12    for every new spam message do
13            tokenize message
14            choose weighting scheme
15            set value for risk scale in between 0 to 1
16            calculate :
17                    •   sum of all input signal, Sₖ, k (stored internally by each DC)
18                    •   Kα, magnitudes of k value,
19                    •   threshold, Tₖ
20                    if Kα > 0 and Kα > Tₖ then,
21                            spam message is tagged as the malicious message; and
22                    else
23                            spam message is tagged as the benign message; and
24                    endif
25                    print spam message with it's tagged label (normal or anomalous)
26    endfor
```

**Fig. 2.** The enhanced pseudo-code of dDCA for measuring the risk level of text spam messages.

have been taken into consideration which referring to the level of output signals: (i) tokenized words with high and medium value as the mature content; and (ii) tokenized words only with high value as the mature content. After the identification of mature tokens, the anomalous conditions is predicted via MCAV calculation. The findings demonstrated in [25] that, by considering both high and medium tokenized words has produced an optimized result in assessing the risk level using DCA.

Extending the aforementioned work in [25], this paper objectively aimed with the following purpose to recognize the factors that dominate the true positive value for risk classification. These factors then become the features in considering the design of the tool in assessing risk which addressing the primary objective of this study, to verify the capability of Danger Theory in assessing risk for text spam messages. Furthermore, an additional simulation using the established non-immune classifiers (Support Vector Machine, SVM and Naïve Bayesian, NB) is also executed to compare the performance with the DCA and dDCA.

1. To distinguish which variants (between DCA and dDCA) are producing a higher true positive rate of risk classification by applying the best parameters value identified from [25] (Experiment 1);
2. To study the impact of antigen multiplication, both on DCA and dDCA. This is vital as to verify the role of signal weight across a different volume of spam or antigen that may cause by antigen deficiency [21]. It is expected that different calculated

weight for input signals may behave differently and affect the risk output values measured from DCA (Experiment 2) and dDCA (Experiment 3); and

3. To compare the performance of immune classifier (DCA and dDCA) with non-immune classifiers (SVM and NB) in assessing the risk level of text spam messages (Experiment 4).

## 4.2   Performance Measurement

All experiments are measured using true positive value as the performance metric for classification rate. True positive rate reflected the value of correct classification for spam messages according to its risk level and context of the spam message.

$$True\ Positive\ (TP) = \frac{\sum messages\ that\ are\ truly\ risk - level\ classified}{\sum All\ messages} \quad (3)$$

The true positive rate is defined in percentage (%), which its total number of truly risk-classified messages is divided with the total number of all messages that have been assessed. All messages are referring to the total number of correctly-classified and falsely-classified spam messages. The higher the value of this TP depict that the better is the performance of the classifier.

## 4.3   Experiments

From initial work executed earlier [25], it is showed that TF is the best-suited term weighting scheme for DCA; TP value is 100% with pre-processing and 84.6% without pre-processing. In addition to that, as tabulated in Table 1, from multiple choices of the risk scale and signal weight matrix, S1 and WM1 are verified as the best-applicable risk scale and signal weights matrix employed using DCA. Risk scale is applied to map the calculated term weight value as input signals and also to map the value of output signals to identify its associated risk level.

Experiments 1–3 are conducted using the above-identified parameters. Although TF is validated as the best-suited weighting schemes, all three terms weighting schemes are continually employed in this work to varying the results of a wider comparison and analysis.

**Table 1.**  Risk scale, S1 and signal weights matrix, WM1 that is identified as best-applicable to be employed together with DCA.

| Risk scale (S1) | | | Signal weights matrix (WM1) | | | |
|---|---|---|---|---|---|---|
| Range | Input signals | Output signals | Input signals | PAMPs | Danger | Safe |
| 1.00–0.70 | PAMPs | High | CSM | 1.0 | 0.5 | 1.5 |
| 0.69–0.40 | Danger | Medium | smDC | 0.0 | 0.0 | 1.0 |
| 0.39–0.00 | Safe | Low | mDC | 1.0 | 0.5 | 1.5 |

# 5  Results and Analysis

## 5.1  Experiment 1 – DCA Versus DDCA

Comparable analysis between DCA and dDCA and the result of the true positive rate for risk classification is tabulated as follows:

From Table 2 and Fig. 3, it has shown that dDCA produced higher true positive value (both with and without pre-processing) than DCA in assessing the risk level. However, for mapping the output signal value to risk scale in determining the risk level, there is a slightly different compared to DCA. For output signal, value 0.39–0.00 is tagged as Medium risk level since; the Low-risk level is a negative value (below 0) in dDCA. The details of the results can be found in Appendix 1.

**Table 2.** Classification True Positive (TP) rate between DCA and dDCA.

| Weighting schemes | DCA (High & Medium for MCAV) | | dDCA ($K_\alpha$) | |
|---|---|---|---|---|
| | With pre-processing | Without pre-processing | With pre-processing | Without pre-processing |
| TF | 100 | 84.6 | 100 | 100 |
| IG ratio | 69.2 | 61.5 | 76.9 | 84.6 |
| $CHI^2$ | 30.8 | 15.4 | 69.2 | 23.1 |

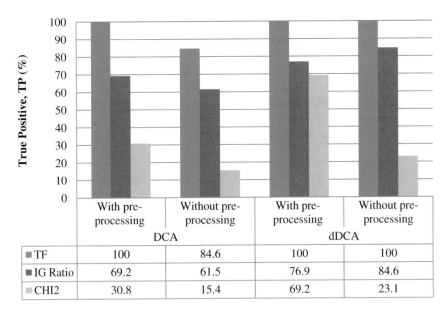

**Fig. 3.** DCA versus dDCA: Risk classification, TP of a text spam message according to term weighting schemes with and without pre-processing.

## 5.2    Experiment 2 – The Effect of Antigen Multiplication Using DCA

The testing is further executed with the falsely classified message (MCAV considered both high and medium value as a mature context) and tested using antigen multiplier with the factor of 10, 40 and 100 times. This Experiment 2 is using DCA as the classifier.

For the experiment with pre-processing, it is not repeated with TF. This is due to the previous results from Experiment 1 that TF has 100% true positive classification rate. However, it is repeated for experiment without pre-processing since there is a false-classified message (see Table 2, TF with DCA).

As tabulated and depicted in Table 3 and Fig. 4, it has shown that antigen multiplication has its effect on DCA in increasing the true positive value. This is obvious for IG Ratio scheme in assessing the risk, especially with multiplication at 40 times. However, there is no significant improvement demonstrated by $CHI^2$ scheme in this experiment, although the multiplication is up to 100 times. This result somehow depicted that $CHI^2$ is not a suitable scheme to derive the precise value of input signals for measuring the risk.

**Table 3.** Classification True Positive (TP) rate with an impact of antigen multiplier for DCA.

| Weighting schemes | Antigen multiplier with pre-processing | | | Antigen multiplier without pre-processing | | |
|---|---|---|---|---|---|---|
| | 10x | 40x | 100x | 10x | 40x | 100x |
| TF | -na- | -na- | -na- | 50 | 50 | 50 |
| IG Ratio | 0 | 100 | 100 | 0 | 80 | 60 |
| $CHI^2$ | 0 | 0 | 0 | 0 | 0 | 0 |

## 5.3    Experiment 3 – the Effect of Antigen Multiplication Using DDCA

In Experiment 3, all messages that falsely classified in Experiment 1 are further tested with the antigen multiplication with the factor of 10, 40 and 100 times using dDCA as the classifier. Same as DCA, as for initial population, all 747 spam messages have been multiplied by 10, 40 or 100 times before it is deployed into RapidMiner for calculation of its input signals using term weighting schemes (TF, IG Ratio, and $CHI^2$). Then, the risk value associated with the message is assessed using dDCA to produce output signals.

This experiment is not repeated for TF due to the previous results (see Table 2, TF with dDCA) that TF has 100% true positive classification rate for both with and without pre-processing.

From Table 4 and Fig. 5, it has shown that the antigen multiplication has its effect for dDCA in increasing the true positive value for assessing the risk, especially for IG Ratio scheme with multiplication at 40 times. However, there is no significant improvement demonstrated by $CHI^2$ scheme in this experiment, although the

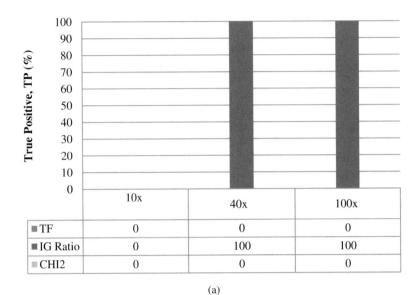

(a)

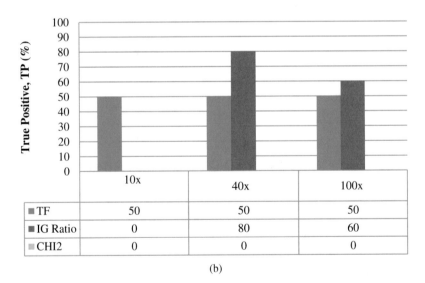

(b)

**Fig. 4.** DCA: Risk classification, TP of a text spam message for 10, 40, and 100 times of antigen multiplication (with pre-processing (a) and without pre-processing (b)).

**Table 4.** Classification True Positive (TP) rate with an impact of antigen multiplier for dDCA.

| Weighting schemes | Antigen multiplier with pre-processing | | | Antigen multiplier without pre-processing | | |
|---|---|---|---|---|---|---|
| | 10x | 40x | 100x | 10x | 40x | 100x |
| IG Ratio | 0 | 100 | 100 | 50 | 50 | 50 |
| CHI$^2$ | 0 | 0 | 0 | 0 | 0 | 0 |

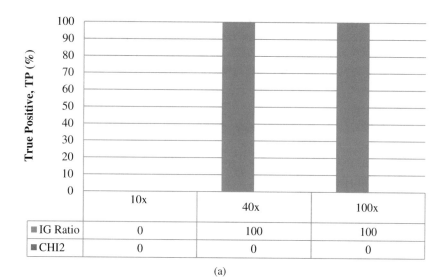

(a)

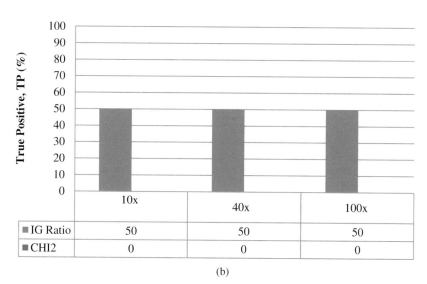

(b)

**Fig. 5.** dDCA: Risk classification, TP of a text spam message for 10, 40, and 100 times of antigen multiplication (with pre-processing (a) and without pre-processing (b)).

multiplication is up to 100 times. This result somehow depicted that $CHI^2$ is not a suitable scheme to derive the precise value of input signals for measuring the risk.

## 5.4    Experiment 4 – Non-immune Classifier (SVM and NB) Versus Immune Classifier (DCA and DDCA) in Assessing the Risk of Text Spam Messages

From this experiment, it is shown that DCA and dDCA classifier outperforms better in assessing the risk of text spam messages than SVM and NB classifier.

From a graph depicted in Fig. 6, DCA and dDCA classifier demonstrated a better rate of true positive (100%) in assessing the risk level of the spam message, compared to SVM (84.62%) and NB (69.23%). For SVM and NB, the risk level is determined by mapping the confidence value of spam with the pre-set risk scale, S1 as tabulated in Table 1. The details of the results can be found in Appendix 2.

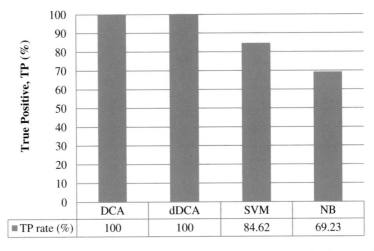

| | DCA | dDCA | SVM | NB |
|---|---|---|---|---|
| ■ TP rate (%) | 100 | 100 | 84.62 | 69.23 |

**Fig. 6.** Comparable analysis for risk assessment of text spam messages using immune classifier (DCA and dDCA) versus with non-immune classifier (SVM and NB).

## 5.5    Discussion

Throughout the proof of concept derived from the whole experiments, analysis of the findings is observed as the following:

- The calculation for measuring risk of text spam messages using dDCA is simpler than DCA due to less parameter that needs to be considered. In dDCA, the only parameter of $K_\alpha$ value needs to be inspected in order to decide the anomaly magnitude of the spam message, whereas in DCA there are two parameters that are required to be equivalent; output signals (smDC, mDC) and MCAV. This has been elaborated in [31] that the dDCA version has been created by removing less important or unnecessary components for the ease of analysis. In addition to that, dDCA does not require additional signal weights to transform input signals into output signals, and fewer parameters are being considered for the calculation. Hence, this is possibly able to reduce the computational cost (for automated method);

- The true positive rate for classifying the risk level in the simulation with pre-processing phase consistently outperformed the experiment that is conducted without pre-processing. This result has enhanced and reflected the claim of finding in [23] that pre-processing has its significant effect in increasing the accuracy rate of classification;

- Antigen multiplication revealed that it is able to contribute its associated value to the creation of input signals. The saturated value is the factor of 40 times of antigen multiplication with TF and IG Ratio. However, the findings of experiments also reflected that $CHI^2$ demonstrated that there is no significant result when integrated together with the antigen multiplication, both with DCA and dDCA;

- In this domain of research, it is mandatory to choose term weighting scheme (as feature selection method) that consider the higher the value of the term, the more relevant the term is in spam classification. The derived value then will fit the proposed risk scale and the theory of anomaly metrics (MCAV and $K_\alpha$). It is also beneficial to consider the value which is normalized in between 0 to 1, the closer the value to 1, the more it is considered as malignant; and

- Throughout the simulation, $CHI^2$ is verified as not a suitable feature selection method (term weighting scheme) to be employed for this domain of task. $CHI^2$ only deliver inhibitory signals (low signal value) which contradict with DCA and dDCA that require at least two types of signal, activating (danger) and inhibitory (safe) signals to assess the risk level. The findings from this experiment somehow suggested that $CHI^2$ signal value tends to generate unreliable and indecisive classification results.

The result of these simulations supports the idea that Danger Theory may become a suitable and feasible solution due to its unsupervised learning paradigm and low weight in computation [32]. Moreover, the simplified algorithm version of this theory verified that dDCA is better and outperforms DCA.

# 6  Conclusion

The evolution of less-featured mobile phone to a smartphone has increased the spam's risk impact since its ability has attached to the Internet technology advancement. In addition to that, unawareness about security implication by users has worsened the problematical situation and many interesting results indicating the potential of losses have been reported. Previous studies have primarily concentrated on distinguishing between ham and spam messages, and less likely for assessment of a spam risk level.

Stimulated by this issue, an implicit decision maker is proposed to be constructed, methodologically inspired by artificial of the human body immune system. Throughout these experiments, an employment of Danger Theory variants verified that dDCA is consistently outperformed than DCA in terms of true positive rate for classification of risk intensity. In addition to that, risk concentration calculated using dDCA is finer-grained compared to DCA. Furthermore, it is also verified that this immune classifier (DCA and dDCA) performed better when it is compared to the non-immune classifier (SVM and NB) that are commonly known as the most feasible algorithm in spam detection. This simulation has demonstrated that Danger Theory is a reliable immune theory for application in risk measurement.

More broadly, an automated environment for conducting these experiments is extremely required. A prototype for both DCA and dDCA needs to be developed and a larger size of the corpus will be deployed to verify its consistency and reliability, as that has been claimed in this paper. At the same time, as to widen the usage of the future developed prototype; a further simulation is suggested to validate the assessment for severity concentration in another form of text spam messages such as Twitter and email messages.

In addition to that, via this testing, a significant role of text mining has been identified. In this case of risk measurement, pre-processing has reduced the high dimensionality data or noises that commonly cause deficiency in retrieving the information. Other than that, the employment of term weighting scheme and risk scale has assisted the classifier to distinguish different levels of input and output signals.

Implications of the results and future research directions have been presented in this paper. Finally, this study may facilitate improvements in the information retrieval task and possibly produce remarkable contributions to knowledge discovery especially in the field of text mining.

**Acknowledgments.** This research is fully funded by the Ministry of Higher Education of Malaysia and Research Management Centre of USIM via grant research with code USIM/FRGS/FST/32/50315.

# Appendix 1

**Experiment 1.** Characteristics of the testing setup for results tabulated in Table 5:

- Term weighting scheme: Term Frequency (TF).
- The value for risk scale (S1) and signal weight matrix (WM1) as tabulated in Table 1.
- For MCAV calculation, mature content referred to both High and Medium tokens.
- Text pre-processing is applied and antigen multiplication is not applied.
- See Sect. 5: Results and Analysis, Experiment 1; Table 2 and Fig. 3 for tabulated results and graph.

According to the risk concentration value tabulated in Table 5, it is empirically proven that DCA and dDCA are feasible algorithms that able to produce a risk level classification for text spam messages. The combination of more than one risky term may result in high risk such as containing *URL* that requested users to respond accordingly with the content of the message. Via this simulation, it is demonstrated that

**Table 5.** Measurement of risk concentration for text spam messages using DCA and dDCA.

| Text D no. | Content of the spam messages | Measured risk level using DCA | | | Measured risk level using dDCA | | |
|---|---|---|---|---|---|---|---|
| | | Output signal | Risk value | Risk level | Output signal | Risk value | Risk level |
| 614.txt | Wanna have a laugh? Try CHIT-CHAT on your mobile now! Logon by txting the word: CHAT and send it to No: 8883 CM PO Box 4217 London W1A 6ZF 16 + 118p/msg rcvd | mDC > smDC | 0.83 | High | $K_\alpha > T_k$ | 0.63 | Medium |
| 712.txt | If you don't, your prize will go to another customer. T&C at www.t-c.biz 18 + 150p/min Polo Ltd Suite 373 London W1 J 6HL Please call back if busy | mDC > smDC | 1.00 | High | $K_\alpha > T_k$ | 0.81 | High |
| 509.txt | Congratulations U can claim 2 VIP row A Tickets 2 C Blu in concert in November or Blu gift guaranteed Call 09061104276 to claim TS&Cs www.smsco.net costÂ£3.75max | mDC > smDC | 0.89 | High | $K_\alpha > T_k$ | 0.72 | High |
| 75.txt | Your credits have been topped up for http://www.bubbletext.com Your renewal Pin is tgxxrz | mDC > smDC | 1.00 | High | $K_\alpha > T_k$ | 0.86 | High |
| 181.txt | You have 1 new voicemail. Please call 08719181503 | mDC > smDC | 1.00 | High | $K_\alpha > T_k$ | 0.78 | High |
| 230.txt | 500 free text msgs. Just text ok to 80488 and we'll credit your account | mDC > smDC | 1.00 | High | $K_\alpha > T_k$ | 0.61 | Medium |
| 36.txt | Text & meet someone sexy today. U can find a date or even flirt its up to U. Join 4 just 10p. REPLY with NAME & AGE e.g. Sam 25. 18 -msg recd@thirtyeight pence | mDC > smDC | 1.00 | High | $K_\alpha > T_k$ | 0.57 | Medium |
| 533.txt | Bored housewives! Chat n date now! 0871750.77.11! BT-national rate 10p/min only from landlines! | mDC > smDC | 0.75 | High | $K_\alpha > T_k$ | 0.52 | Medium |
| 15.txt | Did you hear about the new "Divorce Barbie"? It comes with all of Ken's stuff! | mDC > smDC | 0.00 | Low | $K_{\alpha<}T_k$ | -0.11 | Low |
| 111.txt | Romantic Paris. 2 nights, 2 flights from Â£79 Book now 4 next year. Call 08704439680Ts&Cs apply | mDC > smDC | 0.67 | Medium | $K_\alpha > T_k$ | 0.40 | Medium |
| S1.txt | RM0.00. FREE RM15 k travel voucher! Own a spacious Townvilla with ZERO entry cost & EARN up to RM20 k for home owner programme. Call 0123568311 for details.*T&C apply | mDC > smDC | 0.83 | High | $K_\alpha > T_k$ | 0.54 | Medium |

*(continued)*

**Table 5.** (*continued*)

| Text D no. | Content of the spam messages | Measured risk level using DCA | | | Measured risk level using dDCA | | |
|---|---|---|---|---|---|---|---|
| | | Output signal | Risk value | Risk level | Output signal | Risk value | Risk level |
| S2.txt | ZI6036S Bonus code RM10 ncity888.com http://goo.gl/ eR0CI5 12 win Newtown SCR 888 Sun city 100% Fast Withdraw Whatapps 0182546092 Wechat: csncity8 | mDC > smDC | 1.00 | High | $K_\alpha > T_k$ | 0.94 | High |
| S3.txt | Congratulations, you have successfully activated your TM Rewards membership account. To log into TM Rewards, go to www. tm.com.my/tmrewards | mDC > smDC | 1.00 | High | $K_\alpha > T_k$ | 0.59 | Medium |
| True positive (%) | | **100** | | | **100** | | |

messages that contain information that requested users to respond (*call, text,* and *chat*) are in the appropriate high/medium risk category. While for messages that contain no requirement for users to respond (for instance a message with ID 15.txt) produced low-risk concentration level or potentially to be considered as a non-spam message.

# Appendix 2

**Experiment 4.** Characteristics of the testing setup for results tabulated in Table 6:

- The value for risk scale (S1) as tabulated in Table 1.
- Text pre-processing is applied.
- See Sect. 5: Results and Analysis, Experiment 4 and Fig. 6 for tabulated results and graph.
- *Italic and underlined font* indicates the falsely-classified for risk level

The value of confidence for the non-immune classifier is derived from the data mining tool, RapidMiner. This confidence value is referring to the probability of the message is tagged as spam. The risk level measured for the non-immune classifier is determined by mapping the confidence value of spam with the pre-set risk scale, S1.

**Table 6.** Measurement of risk concentration for text spam messages using immune classifier and non-immune classifier.

| Text ID no. | Measured risk level using immune classifier | | | | Measured risk level using non-immune classifier | | | |
| --- | --- | --- | --- | --- | --- | --- | --- | --- |
| | DCA | | dDCA | | SVM | | NB | |
| | Risk value | Risk level | Risk value | Risk level | Confidence of spam | Risk level | Confidence of spam | Risk level |
| 614.txt | 0.83 | High | 0.63 | Medium | 0.79 | High | 1.00 | High |
| 712.txt | 1.00 | High | 0.81 | High | 0.75 | High | 1.00 | High |
| 509.txt | 0.89 | High | 0.72 | High | 0.77 | High | 1.00 | High |
| 75.txt | 1.00 | High | 0.86 | High | 0.77 | High | 1.00 | High |
| 181.txt | 1.00 | High | 0.78 | High | 0.44 | Medium | 1.00 | High |
| 230.txt | 1.00 | High | 0.61 | Medium | 0.41 | Medium | 1.00 | High |
| 36.txt | 1.00 | High | 0.57 | Medium | 0.77 | High | 1.00 | High |
| 533.txt | 0.75 | High | 0.52 | Medium | 0.75 | High | 1.00 | High |
| 15.txt | 0.00 | Low | −0.11 | Low | *0.65* | *Medium* | *1.00* | *High* |
| 111.txt | 0.67 | Medium | 0.40 | Medium | 0.59 | Medium | 1.00 | High |
| S1.txt | 0.83 | High | 0.54 | Medium | 0.51 | Medium | *0.00* | *Low* |
| S2.txt | 1.00 | High | 0.94 | High | 0.58 | Medium | *0.00* | *Low* |
| S3.txt | 1.00 | High | 0.59 | Medium | *0.35* | *Low* | *0.00* | *Low* |
| True positive (%) | **100** | | **100** | | **84.62** | | **69.23** | |

# References

1. Bujang, Y.R., Hussin, H.: Should we be concerned with spam emails ? A look at its impacts and implications. International Islamic University Malaysia
2. Theoharidou, M., Mylonas, A., Gritzalis, D.: A Risk Assessment Method for Smartphones (2016)
3. Zhang, Y., Xiao, Y., Ghaboosi, K., Zhang, J., Deng, H.: A survey of cyber crimes. Secur. Commun. Netw. **5**, 422–437 (2011)
4. de Natris, W.: Best Practice Forum on Regulation and Mitigation of Unsolicited Communications (2014)
5. Yeboah-Boateng, E.O., Amanor, P.M.: Phishing, SMiShing & Vishing: an assessment of threats against mobile devices. J. Emerg. Trends Comput. Inf. Sci. **5**(4), 297–307 (2014)
6. Zainal, K., Jali, M.Z.: A perception model of spam risk assessment inspired by danger theory of artificial immune systems. In: International Conference on Computer Science and Computational Intelligence (ICCSCI), vol. 59, pp. 152–161 (2015)
7. Timmis, J., Knight, T., de Castro, L.N., Hart, E.: An Overview of Artificial Immune Systems (2002)
8. Liu, F., Wang, Q., Gao, X.: Survey of Artificial Immune System, pp. 985–989 (2005)
9. Dasgupta, D.: Advances in artificial immune systems. IEEE Comput. Intell. Mag. **1**(4), 40–49 (2006)
10. Hart, E., Timmis, J.: Application areas of AIS: the past, the present and the future. Appl. Soft Comput. J. **8**(1), 191–201 (2008)

11. Dasgupta, D., Yu, S., Nino, F.: Recent advances in artificial immune systems: models and applications. Appl. Soft Comput. J. **11**(2), 1574–1587 (2011)
12. Read, M., Andrews, P., Timmis, J.: Artificial Immune Systems (2008)
13. Matzinger, P.: Tolerance, danger and the extended family. Annu. Rev. Immunol. **12**, 991–1045 (1994)
14. Greensmith, J.: The Dendritic Cell Algorithm. University of Nottingham (2007)
15. Greensmith, J., Aickelin, U.: The Deterministic Dendritic Cell Algorithm (2008)
16. Brownlee, J.: Dendritic cell algorithm. In: Clever Algorithms: Nature Inspired Programming Recipes. Creative Commons, pp. 312–318 (2011)
17. Greensmith, J., Aickelin, U.: Artificial dendritic cells: multi-faceted perspectives. In: Human-Centric Information Processing Through Granular Modelling, vol. 182, pp. 375–395 (2009)
18. Greensmith, J., Aickelin, U., Cayzer, S.: Detecting Danger : The Dendritic Cell Algorithm (2010)
19. Aickelin, U., Greensmith, J.: Sensing danger: innate immunology for intrusion detection. Inf. Secur. Tech. Rep. **12**(4), 218–227 (2007)
20. Greensmith, J., Aickelin, U., Twycross, J.: Articulation and clarification of the dendritic cell algorithm dendritic cells (2009)
21. Gu, F., Greensmith, J., Aickelin, U.: Further exploration of the dendritic cell algorithm: antigen multiplier and time windows. In: 7th International Conference Artificial Immune System, pp. 142–153 (2008)
22. Musselle, C.J.: Insights into the Antigen Sampling Component of the Dendritic Cell Algorithm (2010)
23. Zainal, K., Jali, M.Z.: A review of feature extraction optimization in SMS spam messages classification. In: International Conference on Soft Computing in Data Science (SCDS), vol. 545, pp. 158–170 (2016)
24. Zainal, K., Jali, M.Z.: The design and development of spam risk assessment prototype. in silico of danger theory variants. Int. J. Adv. Comput. Sci. Appl. **8**(4), 401–410 (2017)
25. Zainal, K., Jali, M.Z.: The significant effect of feature selection methods in spam risk assessment using dendritic cell algorithm. In: International Conference on Information and Communication Technology (ICoICT 2017), pp. 277–284 (2017)
26. Sethi, G., Bhootna, V.: SMS spam filtering application using Android. Int. J. Comput. Sci. Inf. Technol. **5**(3), 4624–4626 (2014)
27. Zhang, H., Wang, W.: Application of Bayesian method to spam SMS filtering. In: IEEE, pp. 1–3 (2009)
28. Uysal, K., Gunal, S., Ergin, S., Gunal, E.S.: The impact of feature extraction and selection on SMS spam filtering. IEEE **19**(5), 67–72 (2013)
29. Uysal, A.K., Gunal, S., Ergin, S., Gunal, E.S.: A novel framework for SMS spam filtering. IEEE (2012)
30. Almeida, T.A., Hidalgo, J.M.G.: UCI machine learning repository (2012). http://archive.ics.uci.edu/ml/datasets/SMS+Spam+Collection#. Accessed 27 Mar 2014
31. Gu, F., Greensmith, J., Aickelin, U.: Theoretical formulation and analysis of the deterministic dendritic cell algorithm. BioSystems **111**(2), 127–135 (2013)
32. Gu, F., Greensmith, J., Aicklein, U.: The dendritic cell algorithm for intrusion detection. In: Biologically Inspired Networking and Sensing: Algorithms and Architectures, Bio-Inspired Communication and Networking, IGI Global, pp. 84–102, January 2011

# Analyzing User Behaviors: A Study of Tips in Foursquare

Nafla Alrumayyan(✉)⬤, Sumayah Bawazeer⬤,
Rehab AlJurayyad⬤, and Muna Al-Razgan⬤

Department of Information Technology, King Saud University,
Riyadh, Kingdom of Saudi Arabia
naflaru@gmail.com, sumayah.bawazeer@gmail.com,
rehabaljurayyad@gmail.com, malrazgan@ksu.edu.sa

**Abstract.** Foursquare is a popular Location Based Social Network (LBSN). It has become a major platform that enables users to share their opinions on locations they have visited through check-ins and writing tips. The massive amount of data generated by Foursquare provides unexpected opportunities to analyze and obtain interesting insights into people and places. Most of the previous research addressed the interesting findings regarding user behavior through check-ins, but not the characteristics of the most visited venues, which we address in our paper. We also analyze sentiment of Arabic text in LBSNs, focusing on Saudi Arabia. We collected data of more than 1000 venues, 50,000 check-ins and 12,000 tips to investigate the different aspects of those venues with low rating and positive comments by our proposed algorithm using sentiment analysis on Arabic tips. More interestingly, we discovered different communities in Saudi Arabia by applying the Latent Dirichlet Allocation (LDA) model as one of the of topic model approaches. We concluded that some venues with low ratings have more visitors due to the range of services available in the region. In addition, the high number of positive tips proves that certain people influence the others' opinions regardless of the restaurant's rating. The LDA model produces latent collections of people with similar interests as communities which indicates their behavior and patterns.

**Keywords:** LBSNs · Foursquare · Arabic sentiment analysis · LDA

## 1 Introduction

We live in an era where social networks are growing in an astonishing way and allowing people more freedom to spread their words, opinions and interests. In addition, with the emergence of Location Based Social Networks (LBSNs), people started to share their locations and interact with each other. Location-based social networks (LBSNs) are mainly depend on shared locations and opinions to give the users a better experience. LBSNs helps users to explore places near their location or in a certain area. Foursquare, one of the most popular Location Based Social Networks (LBSNs), enables users to search for and find the perfect places based on others' opinions, check-ins, ratings and comments. Moreover, it provides several features such as search

© Springer International Publishing AG, part of Springer Nature 2018
M. Alenezi and B. Qureshi (Eds.): *5th International Symposium*
*on Data Mining Applications*, pp. 153–168, 2018.
https://doi.org/10.1007/978-3-319-78753-4_12

places, and explore other people's recommendations [1]. Since Foursquare was launched in 2009, the amount of users, check-ins and venues information has increased rapidly. In September 2015, there were more than 50 million registered users with 10 billion check-ins [2] and by September 2017 the number of check-ins had increased to 12 billion [3]. Foursquare works internationally in more than 110 countries [4].

In Foursquare there are three basic terms which are important to understand. "Check-in" describes when the user has physically visited (or checked in to) a venue. A venue is a business, physical location, or residence where Foursquare users can check-in and share it with others. A "Tip" allows users to share recommendations, warnings, and any other information in order to make their friends' visit to the venue better. Most of previous research analyzed the user behavior and mobility in Four-square through check-ins [5, 6]. However, there are a few studies that use sentiment analysis on the reviews [7–9]. Furthermore, no studies mentioned the characteristics of the most visited venues. Moreover, up to our knowledge, there is no study that analyzes the sentiment of Arabic text in LBSNs.

To address the gap in the research, we provide a detailed analysis of the users' check-ins, comments and ratings in Foursquare to understand the diverse patterns of people and help answer questions such as: why are certain venues frequently visited in Saudi Arabia? What characteristics do they have? Are there venues with low ratings and a high number of visitors? We examined food category in Foursquare because people like to share their experiences with food by posting comments [10]. This is achieved by investigating and analyzing large-scale of data of thousands of places across Riyadh city in the Kingdom of Saudi Arabia. We considered an available approach to define groups of people with similar interests, which were characterized by the places they prefer. This was gained from the Topic model approach, we utilized the Latent Dirichlet Allocation (LDA) [11] to cluster people based on the restaurants they visited and the tips they wrote. The resulting clusters provide an insight of hidden social communities which are representations of a latent behavior of users.

In the following sections, we first present the research regarding location-based social networks (LBSNs). We then describe our data set in Sect. 3. Following this, we describe our methodology for applying sentiment analysis on tips and how we utilize LDA algorithms to detect different communities. Then we present our results in Sect. 4. We conclude the study in Sect. 5.

## 2    Literature Review

Accessing social networks has increased dramatically and hence produced massive amount of content. In this section, we review the related studies that investigate LBSNs. It is divided into four sections: LBSNs Studies, Sentiment Analysis in LBSNs, Foursquare tips and rating, and community detection in LBSNs.

### 2.1    Location-Based Social Networks (LBSN) Studies

LBSN has been addressed in several research. For example, Colombo et al. [5] analyzed the check-ins of users in foursquare to indicate the human mobility and behavior.

Their aim was to improve recommendation systems. They collected the check-ins of venues of two UK cities. The authors found that most of the people like to visit the same places frequently because their social friends influenced them. In the same line, Daniel et al. [6] studied the pattern of check-in in LBSNs to indicate user behavior. The aim of this study was to understand the pattern of human mobility by investigating behavior of 10,000 frequent users in foursquare. The authors discovered different patterns of venue category usage across time scales. Moreover, they contributed to two applications of the user-centered view. First, clustering users based on their behavior. Then resolve the problem of predicting the next movements by creating simple frequency model. The author found that this model outperforms the Markov Models because it used a better representation of time and of temporal regularities. Thus, will lead a significance performance with future prediction.

On the other hand, Li et al. [10] studied the popular venues in Foursquare by analyzing their common characteristics to discover why these venues attract more visitors. They collected 2.4 million venues from 14 countries using Foursquare search API. The authors observed three aspects that influence the popularity of venues: venue profile information, venue category, and venue age. The results showed that venues with complete profile information were more popular. Whereas, venues under Food category have the highest number of tips generated by people.

## 2.2 Sentiment Analysis in LBSNs

There are a very few numbers of studies about sentiment analysis in LBSNs to classify positive and negative opinions. Chen et al. [7] studied different aspects of Foursquare tips in order to help people to find out about good places. The authors collected 6.52 million of tips by launching 60 virtual machines in US data center using Microsoft azure platform. They conducted sentiment analysis and came up with the concept of "happiness index" to know the group of users that they post positive tips. In addition, they discovered the venues that could make people happier. Moreover, Gallegos et al. [8] investigated the features of areas in a city that people tend to be more delighted when they visited them. The dataset collected from foursquare includes the venues located at US metropolitan areas and have many check-ins. They also extracted tweets relevant to Foursquare in order to analyze the properties of the places. The authors applied sentiment analysis using SentiStrength tool to extract positive and negative feeling. They analyzed the tweets shared from different locations and observed that the venues shared through twitter with a lot of check-ins tend to have positive tips. Their result indicated that the areas that have various activities and places such as restaurants, gyms, and beaches attract more visitors.

## 2.3 Foursquare Tips and Rating

Reviews and rating are available for each specific location in LBSNs. It gave the participants (users) a clear insight about the places and might be a significant factor influencing their decisions.

Hajas et al. [12] analyzed the reviews and ratings of the Yelp users over a period of 7 years. The authors focused on neighbor restaurants of 20 universities in the US. They provided a model to interpret the behavior of Yelp ratings and its relationship with the quality and services of the restaurant. Moreover, they studied the cyclic behavior of the most reviewed restaurants. They discovered the cumulative ratings for a period will converge, in contrast to the reviews of individual restaurants, and restaurants close to each other received frequent ratings.

Along the same line of Yelp rating, Nabiha [13] analyzed the reviews of the Yelp users to predict the user's star rating of a particular restaurant. The dataset consisted of 14,000 restaurants and 706,646 reviews. Four models were applied to analyze the semantic of the text and to extract useful features. He trained the model based on four supervised learning algorithms: Logistic Regression, Naive Bayes, Perceptron and SVC. The results presented that Logistic Regression along with unigrams & bigrams proved to be the most effective predictive powers. In addition, AlMohanna et al. [14] proposed a novel approach that analyzes Arabic restaurants' reviews to explore the relationship between rating and sentiments' scores of the review text. They investigated their approach on the top 10 restaurants of Saudi Arabia from Qaym website [15]. Sentimental analysis tool and manual coding of the reviews were used to analyze the text. The result has shown the variance ranking of restaurants based on different utility measures.

## 2.4  Community Detection in LBSNs

Community detection in location-based social networks (LBSN) is an interesting topic that has attracted researchers to improve social interactions. Hao et al. [16] presented a detection approach to discover the social evolution of location- focused communities in LBSs. The authors were concerned about detecting the location- focused communities' behavior at particular time and for popular set of users through their check-ins. Ulti-mately, they found that group of communities have similar behavior while considering the time and location.

Wang et al. [17] proposed framework to detect overlapping communities in LBSNs. They came up with a novel approach that able to cluster like-minded users from different social perspectives. The authors conduct an empirical study to evaluate the efficiency of the model. The model was evaluated based on foursquare data with 9,803,764 check-ins worldwide. Afterwards, for each detected community they use LDA model to define the similarities of tips and calculate the average community tips similarities. While others use LDA as model to detect the communities such as [11, 18]. They studied foursquare check-ins in New York to detect different communities via topic modeling technique LDA. They treat the model on geo-spatial location, time, venue categories. Wang et al. [19] proposed a community clustering framework to detect overlapping communities by considering the heterogeneous social interactions in LBSNs. They collected 12 million check-ins performed by almost 720,000 users over 3 million venues from foursquare. They aggregated users and venues that had some similarity. They also conducted an experiment to evaluate the quality of the community's detection to group like-minded users from different social perspectives and granularity.

From the above-mentioned research, one can notice that most of the previous studies focused on analyzing user behavior through check-ins. In addition, up to our knowledge, there is no study mentioned the characteristics of the most visited venues. Also, there are a few studies that used sentiment analysis to reviews text. Thus, our project will try to address this limitation, where we will be analyzing venues and sentiment of Arabic text in LBSNs particularly Saudi Arabia. In our study, we will attempt to uncover the relations between place rating score, place comments, and place check-ins to discover people behaviors and patterns.

## 3   Methodology

This section describes the dataset used in our research and provides an overview of our methodology. Our methodology consists of the following: data collection, sentiment extraction and analysis, and the Latent Dirichlet Allocation (LDA) Model.

### 3.1   Data Collection

We have used Foursquare's Search API in order to collect the venues in specific regions [20]. Since we focused on Saudi, especially the capital, we customized Foursquare's API to only retrieve food venues in Riyadh city. We collected our data during October 2017 and there were approximately 1000 venues. Table 1 shows the list of venues with different food categories in different locations in Riyadh city. Each venue includes the following attributes: venue name, venue ID, address, contact number, restaurant type, venue tips, number of check-ins, venue price tier, number of tips, and number of users. Moreover, the venue includes statistical data such as a number of check-ins, a number of tips and number of users visiting each venue. Our collected data generated over 50,000 check-ins, 12,000 tips, and 6000 users. These three statistics: number of check-ins, number of tips and number of users who visited each venue will help us in detecting the community and understanding people's behavior.

**Table 1.** Sample dataset of venues information extracted from Foursquare.

| venue.name | venue.location.address | venue.categories | venue.checkinsCount | venue.usersCount | venue.tipCount | venue.price.tier |
|---|---|---|---|---|---|---|
| Mirage (ميراج) | Takhassusi St. | Chinese Restaurant | 10506 | 6131 | 347 | Moderate |
| Dunkin' Donuts (دانكن دونتس) | Telecom Mall | Donut Shop | 10489 | 2646 | 51 | Cheap |
| Alâan (الآن) | 280 Orouba Rd. | French Restaurant | 4571 | 2983 | 316 | Moderate |
| Sultana (سلطانة) | Takhassusi St. | Moroccan Restaurant | 2924 | 2414 | 345 | Moderate |
| Sizzlerhouse (سزلر هاوس) | Takhassusi St | Restaurant | 239 | 195 | 22 | Moderate |
| Gulf Royal Chinese Restaurant (مطعم الخليج الصيني) | King Fahd Rd. | Chinese Restaurant | 4863 | 2705 | 144 | Moderate |
| Ristorante La Casa Italiana (البيت الإيطالي) | King Fahd Rd. | Italian Restaurant | 1509 | 1016 | 102 | Moderate |
| Julia's (جوليانز) | Takhassusi St. | Italian Restaurant | 6383 | 4476 | 488 | Expensive |
| Lezwan Café (ليزوان القهوة النجدية) | Arouba Rd | Café | 102 | 54 | 4 | Cheap |
| Golden Saj (الصاج الذهبي) | Takhassusi Street | Middle Eastern Restaurant | 1813 | 1068 | 59 | Moderate |
| Shawirma Badawyah (شاورما بدوية) | Al Takasusi St. | Middle Eastern Restaurant | 155 | 102 | 7 | Moderate |
| Krispy Kreme (كريسبي كريم) | Al Takhassusi St | Donut Shop | 1071 | 667 | 13 | Cheap |

## 3.2    Sentiment Extraction and Analysis

With the emergence of social media and the increase of content, sentiment analysis became an interesting area of research [9]. To the best of our knowledge, this study is the first to perform sentiment analysis on tips written in Arabic regarding Riyadh's restaurants on Foursquare. Furthermore, we will introduce the idea of investigating the influence of people's opinions when the restaurant's rating is low. To achieve our objective, we extracted the users' tips written in Arabic on each retrieved venue in order to analyze people's opinions using Foursquare's API. Next, we analyzed the sentiment of each restaurant using Lexicon-based approach. The lexicon is dedicated to opinions related to food and consists of 734 words and phrases in the Saudi dialect along with their polarity/semantic orientation (positive/negative). Example of popular words in the Saudi dialect that were extracted from Foursquare are: الاكل خايس، صحن المشاوي لا يعلى عليه ، مهب مره اكلهم زيوت. In addition, we extended the lexicon by adding more words and phrases and ended up with more than 2000 words. Furthermore, we added emoticons to our lexicon that represent only one meaning like "😊, ☺, ☻, ◉, 🍴, 👍". Table 2 shows part of our lexicon.

**Table 2.** Lexicon of restaurant opinions in Saudi Dialect.

| Word/Phrase | Sentiment Score |
|---:|---|
| يهبل | Positive |
| المطعم ادماان | Positive |
| مقرف | Negative |
| مره زين | Positive |
| اخس مطعم | Negative |
| خدمتهم خايسه | Negative |
| ماهنا زود | Negative |
| ادمنته | Positive |

As shown in Fig. 1, the first step to classifying venue tips is pre-processing the text by deleting any tip that includes both Arabic and English words. Second, we used ArabicStemR library in R language to segment the sentences into separate tokens [21]. Then, we detected the tokens and emoticons that accurately classify the text. Finally, a sentiment annotation will be assigned to the venue's tip.

We also introduced a new way of analyzing the text of tips. We investigated the correlation between the restaurant's rating and written text (tips). Logically, a low rating indicates a negative restaurant opinion, and a high rating indicates a positive opinion, however, this is not always true. Previous researchers have not addressed this behavior and there are no opposite relationships between rating and (positive/negative) opinion. Therefore, we thought to investigate this issue in details by introducing our approach. Hence, we analyzed the Arabic tips of each restaurant and extracted the sentiment to determine whether the person is happy with the restaurant or not, despite its low rating. Next, we calculated the extent of people's influence. The algorithm in Table 3 shows the pseudocode used to calculate the influence of lowly rated restaurants in detail.

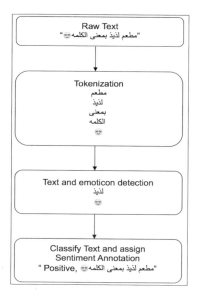

**Fig. 1.** The process of analyzing the sentiment of the text.

**Table 3.** Pseudocode used to calculate the influence of lowly rated restaurants.

| Pseudocode 1 |
| --- |
| 1: **Input:**    $V_r$ Venue Rating, |
|                     $V_t$ the sentiment extracted from Arabic tips of specific venue |
| 2: **Output:** *IR* Influence Rate |
| 3: **Process:** |
| 4: $Positive_{List} = 0$ |
| 5: **if** $V_r < 7$ stars **then** |
| 6:    **if** $V_t$.polarity == positive **then** |
| 6:       $Positive_{List} = Positive_{List} + 1$ |
| 7:    **end if** |
| 8: **end if** |
| 9: $IR = \dfrac{\#positive\_tips}{\#positive_{tips} + \#negative\_tips}$ |
| 10: Return *IR* |

The inputs of the proposed algorithm are $V_r$ and $V_t$. $V_r$ is the venue rating that we extracted from Foursquare. $V_t$ is the sentiment of the Arabic tips that we previously extracted. The process of the algorithm goes as follows: first it checks if $V_r$ venue rating is less than seven then checks if the polarity of the venue $V_t$.polarity is positive or not. If the sentiment of the tip is positive, it will be added to the list Positive$_{List}$. Once all of the tips have been checked for each venue the influence rate IR will be calculated by dividing the number of positive tips by the total number of positive and negative tips and output will be returned.

### 3.3    Latent Dirichlet Allocation (LDA) Model

To detect different communities in our data, we analysed Foursquare users and tried to detect communities based on their favourite restaurants. We apply the idea of Topic Modeling by using Latent Dirichlet Allocation (LDA). LDA is commonly used to interpret the communities, which are representative of different factors. LDA is a latent space model used to simplified large sets of text corpora and uncovers the hidden topics [11, 18]. It is called a word to document model where documents can be described by considering the frequency of the words within them and how these words relate to the various documents. Therefore, so the heaviest words associated with each other will formulate the hidden topics [18]. We reformulated our data to be as a document represents a restaurant, and each ID for a user who wrote the tips on a specific restaurant can be thought as a word in a document to overcome the issue that our data does not revolve around the concept of themes in the text.

Our experiment is based on four attributes defined in Table 4. The researchers would usually identify a fixed number of topics as it is common to test various numbers of T [11]. In our case, after several trials, it was determined that the ideal number of topics was four or five. MinTips is the minimum number of tips regarding a specific restaurant for inclusion in the analysis. to be included in the analysis. We set it to be between 9 and 30 tips per restaurant, otherwise it is excluded. MinIDs is the minimum number of IDs on each topic "word", in our case we will try 3, 5, and 6 IDs. Finally, weight is the acceptable weights for each "word" ID in a topic, which we set at between 0.005 and 0.5. We follow the same way of [11] to set the above-mentioned value for each attribute.

**Table 4.** Parameters and tested values for model specification for LDA.

| | |
|---|---|
| T | 4,5 |
| MinTips | 9–30 |
| MinIDs | 3, 5, 6, 7 |
| Weight | 0.5–0.005 |

LDA was used to cluster the more popular users into one community. Thus, each two or more users' ID associated together in two or more restaurants will be arranged in a cluster. Each cluster represents a community of users who are likely to visit the same restaurants. The analysis of the resulted communities will be discussed in Sect. 4. Figure 2 below was borrowed from [17], added our enhancement to show our idea of community detection.

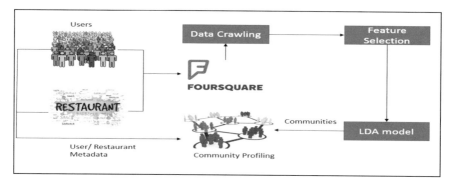

**Fig. 2.** Community detection.

## 4    Results and Discussion

In this section, a detailed analysis of allocated user's activity in Foursquare is illustrated. Moreover, we investigate the different aspects of the venues with only low rate by applying our proposed algorithm in Table 3. Our findings explore varies activity within the course of location, global popularity and food category.

### 4.1    Venue Services and Price

It is remarkably interesting how most people like to visit the expensive restaurant, which has high rates more than cheaper restaurants with similar rate regardless of the distance. In Foursquare, for each venue there are four price categories (very expensive, expensive, moderate and cheap), and a rating that describes the quality of service in a value starting from 0 to 10. Figure 3 shows the number of check-ins and tips for restaurants with a high rating ($r >= 8$) in different price categories. We group both very expensive and expensive as an expensive group for clear illustration purpose.

Figure 3 illustrates two curves present considerably different patterns. Over 10,000 people perform check-ins in expensive restaurants while the cheaper restaurants have a very low number of check-ins. Most of the people in Riyadh choose the restaurant relying on some special characterizes such as if it has an outdoor area, nice decoration, good atmosphere, great service, kind waiters and a sweet menu. Where the cheaper restaurants do not consider these characterize. It also, people think that cheap restaurants anybody can dine-in any time, whereas expensive restaurants not many people visited and like to write their comments and feedback. It also could be people like to show off that they visited these over-priced places.

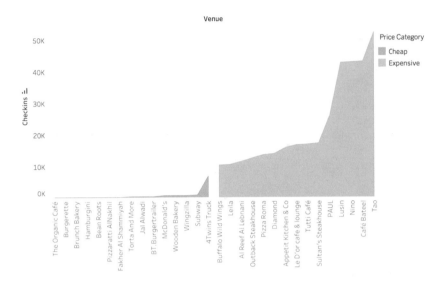

**Fig. 3.** Expensive venues vs cheap venues.

## 4.2    International Venue

We also interested in the impact of international restaurants that have branches outside Saudi Arabia and local people opinion. We plot the venue creation date, a number of check-ins and rating of two international restaurants: Burger Boutique was created in Foursquare on 3rd of May 2010, and recent one Urth Caffe that was created this year on 13 of March 2017. As indicated in the table the rating and number of check-ins, Urth Caffe was recently opened but has high check-ins comparing to Burger Boutique as shown in Table 5. It is observable that people feel confident about the international restaurants based on the idea of "if it is successful internationally it will also be successful locally" [22]. It could be also that Saudi's people like to try American restaurant in terms of their burger and cafe.

**Table 5.** International venues.

| Venue | Creation Data | In 2017 | |
|---|---|---|---|
| | | # Check-ins | Rating |
| Burger Boutique | 3-May-2010 | 72244 | 9.2 |
| Urth Caffe | 13-Mar-2017 | 55405 | 8.8 |

## 4.3    Venues Category

Additionally, we look into a type of restaurants that popular among people in Saudi. Figure 4 plots the number of check-ins across eleven most popular food categories in Foursquare. These groups are (American, Asian, Chinese, Fast Food, Indian, Italian,

Japanese, Mexican, Middle Eastern, Seafood and Turkish). The most interesting result is that "Turkish" restaurants received a high number of check-ins. This could be the similar taste of the culture of food to "Middle Eastern" restaurants as main dishes of shawarma and kabab. Next, is "Fast Food" comes as a favorite type since in modern and busy world many people seek to save time to eat, also could be that in Saudi the number of young people is higher compared to old people [23] and young people tends to like fast food and enjoy it more. On the opposite side is the "Italian", "Seafood", and "Indian" received a fewer number of check-ins. Even though Italian foods are favorable around the world, it does not seem to be of high interest to Saudi cultural. Also, seafood is not that famous, could be that Riyadh is in the middle of a desert, people used to eat red meat more than seafood and also seafood is not fresh. These factors might be different to the area next to the beach as of Jeddah and Dammam. Furthermore, Arab food tends to be mild and this could be one of the reasons why hot and spicy food as of Indian restaurant is not well-liked.

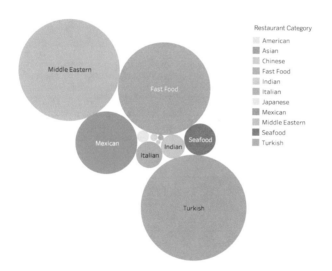

**Fig. 4.** Food categories.

## 4.4   Venue Rating and Positive Sentiment

In this section, we analyze the Arabic tips and comments of popular places in Riyadh, as this is considered one of our contributions in this paper. We conduct sentiment analysis of more than 30 venues that have a low rating to discuss the different aspects that make these venues have a high number of positive tips and thus a high number of visitors, yet low rating. We start by extracting and analyzing the sentiment of the tips for each venue that has rating lower than seven. Then we apply our pseudocode in Table 3 to calculate the influence rate and find out if the people opinions matter most than the venue rating. Figure 5 shows a sample of venues with rating less than seven. As we can notice the number of check-ins (in orange color) is directly proportional to the positive tips (in gray color) despite, the low rating of the venue (in blue color). The

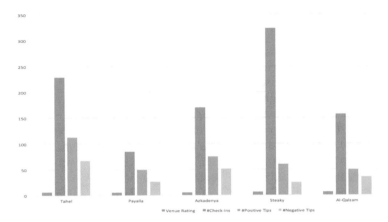

**Fig. 5.** Sample venues with low rating.

**Fig. 6.** Tips word cloud of venues with low rating.

results show that the venues with influence rate more than 50% means the positive opinions surpassing the negative (in yellow color) and have more visitors. Figure 6 shows the frequent words found in the tips of venues with low rating. These five words are the most frequent: لذيذ، الاكل، الجلسات، جميل، مطعم and gave a strong interpretation that the restaurants with a low rating also have positive opinions.

We also investigated the aspects of these venues by understanding why people like to visit such places. Based on the findings, the venues that are located in areas with services, such as a variety of restaurants, activities centers like gyms, and public parks have a number of visitors. Figures 7 and 8 show two of the restaurants that are located in different areas and it is clear these areas have various activities. For instance, "Tahil" restaurant in Fig. 7 surrounded with other restaurants with different type of food, medical center, bookstore, and shopping center. On the contrary, "Cafune" coffee shop as shown in Fig. 8 has high rating than "Tahil" restaurant and there is no variety of places surrounded by it. Thus, the number of visitors is very low as well as most of the comments are negative.

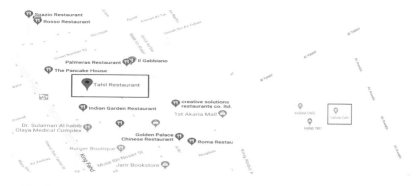

**Fig. 7.** "Tahil" restaurant nearby places          **Fig. 8.** "Cafune" cafe nearby places

## 4.5  Community Detection

In this section, we present the results of running LDA on data from Riyadh. Since Foursquare data does not reveal the real communities, we indirectly sort it detect the communities through the user's ID who create tips on the different restaurants. We detect different communities based on foursquare food category. In our analysis, we examine the "top" restaurants in each community as shown in Table 6. By observing these restaurants, we are able to better understand the users and discover the latent behavior. One can notice that the international restaurant such as IHOP is shared by all of the communities. Moreover, Burger Joint restaurants are the most favored by Saudis. The rating of all communities' restaurants above 7.5, which means the people in Foursquare highly influenced by others' experiences. In addition, moderate prices, expensive and very expensive restaurants are the most visited. However, the traditional restaurants rarely appear between the communities. This is might be because the users in foursquare present themselves as being a specific type of person, they want to show off, or it could be that people will not write tips on frequently visited restaurants as nobody might read it as it popular. There is overlap between the communities as the user may belong to more than one community. Furthermore, we define interesting factors within top restaurants of one cluster, such as users of one of the communities' preferred cheap restaurants, which are also served fast food or coffee. While others preferably more luxury places and services regardless the price. They also favor international food such as Asian and European food. However, most of the users align with the moderate prices and diversification of food.

The figure below shows geo-location of the restaurants using google maps, each color represents one community (Fig. 9).

**Table 6.** Top restaurants of the communities.

| Community | Restaurants | Categories name | Price | Rating |
|---|---|---|---|---|
| 1 | White Garden The Lounge (وايت قاردن ذا لاونج) | Café | Cheap | 8.8 |
| | Gulf Royal Chinese Restaurant (مطعم الخليج الصيني) | Chinese Restaurants | Moderate | 8.5 |
| | Shake Shack (شيك شاك) | Burger Joint | Cheap | 7.8 |
| 2 | Appetit Kitchen & Co (أبيتيت كيتشن) | French Restaurant | Expensive | 8.6 |
| | OREGANO PIZZERIA (اوريجانو بيتزاريا) | Pizza | Moderate | 7.5 |
| | Alaan (آلان) | French Restaurant | Moderate | 8.5 |
| | Boardwalk Burgers & Fries (بوردوك برجر) | Burger Joint | Moderate | 8.1 |
| | Yokari (يوكاري) | Japanese Restaurant | Very Expensive | 8.8 |
| | Burger Boutique ( برجر بوتيك) | Burger Joint | Expensive | 9.2 |
| 4 | Sultana (سلطانة) | Moroccan Restaurant | Moderate | 7.7 |
| | Gulf Royal Chinese Restaurant (مطعم الخليج الصيني) | Chinese Restaurants | Moderate | 8.5 |
| | Armin (آرمين) | Mediterranean Restaurant | Expensive | 8 |
| | Julia's (جولياز) | Italian Restaurant | Expensive | 7.4 |

**Fig. 9.** Geo-location of community's restaurants

## 5 Conclusion and Future Work

In this research, we have study people's patterns in Saudi Arabia using one of the famous LBSNs which is Foursquare. We have shown that the data derived for the capital city of KSA, Riyadh. In particular, our study depends on Food category where people interest to share their experience and leave comments. We provided the aspects of the venues that have a low rating and a large number of visitors by conducting sentiment analysis of venues" Arabic tips. We detect a number of hidden social communities using LDA algorithm based on the concept of topic modeling. Moreover, the resulting communities reveal the latent behavior of Saudis based on their favorite restaurants. Finally, we studied the people in Saudi Arabia by providing a detailed analysis of the users' check-ins, comments, and ratings in Foursquare.

For future work, we will look to extend the study across many cities in Saudi Arabia. Also, we will look more into using other types of information associated with the venues, such as user check-ins data/time, working hours or photos.

## References

1. Foursquare. https://foursquare.com/. Accessed 09 Nov 2017
2. "Foursquare users have checked in over 10 billion times", VentureBeat, 13 September 2016
3. Important foursquare Stats and Facts, November 2017. https://expandedramblings.com/index.php/by-the-numbers-interesting-foursquare-user-stats/
4. How Foursquare hopes to hit profitability. TechCrunch. https://techcrunch.com/2016/05/09/how-foursquare-hopes-to-hit-profitability/. Accessed 09 Dec 2017
5. Colombo, G.B., Chorley, M.J., Williams, M.J., Allen, S.M., Whitaker, R.M.: You are where you eat: Foursquare check-ins as indicators of human mobility and behaviour. In: 2012 IEEE International Conference on Pervasive Computing and Communications Workshops, pp. 217–222 (2012)
6. Preoţiuc-Pietro, D., Cohn, T.: Mining user behaviours: A study of check-in patterns in Location Based Social Networks, May 2017
7. Chen, Y., Yang, Y., Hu, J., Zhuang, C.: Measurement and analysis of tips in foursquare. In: 2016 IEEE International Conference on Pervasive Computing and Communication Workshops (PerCom Workshops), pp. 1–4 (2016)
8. Gallegos, L., Lerman, K., Huang, A., Garcia, D.: Geography of emotion: where in a city are people happier? In: Proceedings of the 25th International Conference Companion on World Wide Web, Republic and Canton of Geneva, Switzerland, pp. 569–574 (2016)
9. Liu, B.: Sentiment analysis and opinion mining. Synth. Lect. Hum. Lang. Technol. 5(1), 1–167 (2012)
10. Li, Y., Steiner, M., Wang, L., Zhang, Z.L., Bao, J.: Exploring venue popularity in Foursquare. In: 2013 Proceedings IEEE INFOCOM, pp. 3357–3362 (2013)
11. Joseph, K., Carley, K.M., Hong, J.I.: Check-ins in 'Blau Space': applying Blau's macrosociological theory to Foursquare check-ins from New York City. ACM Trans. Intell. Syst. Technol. 5(3), 1–22 (2014)
12. Hajas, P., Gutierrez, L., Krishnamoorthy, M.S.: Analysis of Yelp Reviews. ArXiv14071443 Phys., July 2014
13. Asghar, N.: Yelp Dataset Challenge: Review Rating Prediction, May 2016

14. Al Mohanna, N., Al-Khalifa, H.S.: How rational are people? Economic behavior based on sentiment analysis, pp. 124–127 (2014)
15. مطاعم الرياض – قيم. http://www.qaym.com/. Accessed 28 Dec 2017
16. Hao, F., et al.: An efficient approach to understanding social evolution of location-focused online communities in location-based services. Soft Comput. **22**, 1–6 (2017)
17. Wang, Z., Zhang, D., Yang, D., Yu, Z., Zhou, X.: Detecting overlapping communities in location-based social networks. In: Social Informatics, pp. 110–123 (2012)
18. Joseph, K., Tan, C.H., Carley, K.M.: Beyond 'local', 'categories' and 'friends': clustering foursquare users with latent 'topics', p. 919 (2012)
19. Wang, Z., Zhou, X., Zhang, D., Yang, D., Yu, Z.: Cross-domain community detection in heterogeneous social networks. Pers. Ubiquitous Comput. **18**(2), 369–383 (2014)
20. Search for Venues - Foursquare Developer. https://developer.foursquare.com/docs/api/venues/search. Accessed 09 Dec 2017
21. Nielsen, R.: ArabicStemR: Arabic Stemmer for Text Analysis (2017)
22. Bratu, F.: Going Global: How to Succeed in International Markets - wintranslation blog. Professional translation services | wintranslation, 08 December 2011
23. Population aging in Saudi Arabia. http://www.sama.gov.sa. Accessed 20 Dec 2017

# Classification of Customer Tweets Using Big Data Analytics

Abeer Nafel Alharbi[1], Hessah Alnnamlah[1], and Liyakathunisa[2(✉)]

[1] Prince Sultan University, Riyadh, Kingdom of Saudi Arabia
abeer.harbii@gmail.com, hessah.alnamlah@gmail.com
[2] Taibah University, Medina, Kingdom of Saudi Arabia
dr.liyakath@yahoo.com

**Abstract.** Word of mouth has a great impact on commercial planning and decision-making. Social media is considered as one of the greatest media to spread customer's opinion about product. Twitter in particular serves as a platform to share people opinion with the words. Decision makers nowadays are seeking analysis approaches on customer tweets to classify whether a customer is satisfied or unhappy. But the enormous number of tweets per seconds and the live streaming of twitter require big data processors in order to support decision-making. In this paper, we propose a recommender system that helps decision makers to fetch customer streaming tweets and classifies their opinion within seconds. We aim to achieve that by applying Naïve Bayes algorithm using big data machine learning approach, Apache Hadoop and Mahout tools are used. The result of our finding is a recommender system that can be used to classify any new customer tweets. The accuracy of the model is 99.39% which promises accurate results in identifying negative or positive customer opinion about a product in a tweet.

**Keywords:** Big Data Analytics · Recommender system · Mahout · Tweets
Sentiment analysis

## 1 Introduction

Big Data is the buzzword and a number of organizations are there in the markets that are the major sources of Big Data. Social media provides vast amount of data that can be used for further analysis, Twitter is one such source. By applying analytics to structured and unstructured data and mining, that is, extracting items of information which are important for business planning and execution; enterprises are changing the way they plan and make business decisions. Given that many business decisions are influenced by behavioral biases, Sentiment Analysis represents another valuable source of information that aids business decisions and performance evaluation [1]. Sentiment analysis is used to detect and classify sentiments in texts. Usually the sentiment polarity is classified as positive, negative or neutral classes. In this paper, we are going to analyze the sentiment based on the word opinion in the context. We have selected Twitter as a

© Springer International Publishing AG, part of Springer Nature 2018
M. Alenezi and B. Qureshi (Eds.): *5th International Symposium on Data Mining Applications*, pp. 169–180, 2018.
https://doi.org/10.1007/978-3-319-78753-4_13

micro-blogging platform where customers, consumers and users of a product can express their opinion about that product [2].

Diverse types of surveys like product fulfillment survey, competitive products and market survey, brand equity survey, customer service survey, new product acceptance and demand survey, customer trust and loyalty survey and many other surveys for the company and product enhancements has been used by many companies. These kinds of surveys need more efforts, costs high and also consumes more time. The reports generated by this process might not be genuine. For this reason, Online Social Networks (OSNs) such as Face book, Google+, and Twitter has changed the current system in many dimensions. Most of organizations spend lot of money on survey of their products and for getting the feedback, for future enhancements. The main objective of this paper is to reduce the companies' expenses and the time spent on surveys by using the Online Social Networks [2].

Our goal in this paper is to analyze the tweets and other public discussions on the company products by filtering and Natural Language Processing (NLP). Data is analyzed by collecting tweets from twitter about certain products. Further analysis is done on collected data to categorize them into different categories like negative and positive. The final analysis report is generated after the analysis process for the high-end users. The results of our findings help the management to understand the backlogs and the public response on their products and shares. The generated report from this process is genuine compared to the other survey and analysis techniques. Hence the proposed recommender system is highly efficient for all the product-based, service-based and shares related companies. The solution is cost reducing and time saving for customers and the management.

## 2 Related Work

### 2.1 Machine Learning Techniques

In recent years a lot of work has been carried out on sentiment analysis for twitter data using machine learning techniques. A detail survey is provided by Vishal and Sonawane [3].

Pak and Paroubek [4] in 2010 proposed Twitter as a Corpus for Sentiment Analysis and Opinion Mining they showed how to automatically collect a corpus for sentiment analysis and opinion mining purposes. They performed linguistic analysis of the collected corpus and explained discovered phenomena. Using the corpus, they build a sentiment classifier that is able to determine positive, negative and neutral sentiments for a document.

Agarwal et al, [5] in 2011 introduced POS-specific prior polarity features. They explored the use of a tree kernel to obviate the need for tedious feature engineering. They built tree kernel to classify tweets and applied on POS and N-Gram.

Neethu and Rajashree [6] in 2013 proposed Sentiment analysis in twitter using machine learning techniques, they try to analyze the twitter posts about electronic products like mobiles, laptops etc. using Machine Learning approach. They proposed that by doing sentiment analysis in a specific domain, it is possible to identify the effect of domain information in sentiment classification. They present a new feature vector for classifying the tweets as positive, negative and extract peoples' opinion about products.

Geetika Gautam and Divakar Yadav [7] in 2014 proposed Sentiment analysis of twitter data using machine learning approaches and semantic analysis, they first pre-processed the dataset, after that extracted the adjective from the dataset that have some meaning which is called feature vector, then selected the feature vector list and there after applied machine learning based classification algorithms namely: Naive Bayes, Maximum entropy and SVM along with the Semantic Orientation based WordNet which extracts synonyms and similarity for the content feature.

Akshay et al. [8] in 2016, proposed a model for sentiment analysis of tweets with respect to latest reviews of upcoming Bollywood or Hollywood movies. With the help of feature vector and classifiers such as Support vector machine and Naïve Bayes, they classified tweets as positive, negative and neutral to give sentiment of each tweet.

Alec et al. [9] in 2009 proposed Twitter sentiment classification using distant super-vision processing. The authors classified the tweets using the feelings for example posi-tive tweets with positive feelings and negative tweets runs with negative feelings.

### 2.2 Machine Learning Methods

Different machine learning algorithms are used to classify Tweets generated from streaming Internet of Thing Devices (IoT) devices. In our proposed approach we use Naïve Bayes classification. Naive Bayes is a simple model which works well on text classification [10]. For text classification, we use a multinomial Naive Bayes model as shown in Eq. 1, Class c* is assigned to tweet d.

$$C^* \text{argmac}_C P_{NB}(c/d)$$

$$P_{NB}(c/d) = \frac{(P(c)) \sum_{i=1}^{m} P(f/c)^{n_{i(d)}}}{P(d)} \tag{1}$$

Where

$f \rightarrow$ Feature

$n_{i(d)} \rightarrow$ Count of feature ($f_i$) and is present in $d$ which represents a tweet

$m \rightarrow$ Total number of features.

Parameters P(c) and P (f|c) are computed using maximum probability estimates.

### 2.3 Twitter

Twitter is an online person-to-person communication and small scale blogging admin-istration that empowers its clients to send and read instant messages of up to 140 char-acters, known as tweets [11]. Jack Dorsey made twitter in March 2006. The online web-based social networking administration increased overall, with over enrolled clients starting at 2012, creating more than 340 million tweets day by day and more than 1.6 billion inquiry questions for each day, Twitter has turned out to be one of the most wanted sites on the Internet, and has been depicted as "the SMS of the Internet". Unregistered clients can read tweets, while enlisted clients can post tweets through the site interface. There are two sorts of Twitter API available.

**REST API:**  The REST API can be used to make authenticated Twitter API requests.

**Streaming API:**  Associating with the streaming API requires keeping a constant HTTP association open [12]. An application that associates with the streaming APIs won't have the capacity to set up an association in light of a client tasks. Rather, the code for keeping up the streaming association is regularly kept running in a procedure isolated from the procedure that handles HTTP tasks. The streaming procedure gets the info tweets and plays out any examining, separating, or aggregation required before putting away the outcome to an information store. In light of client demands HTTP process the request and stores the information about the results; the advantages of having an ongoing stream of tweet information makes the coordination valuable for some sorts of uses [12].

### 2.4    Big Data Analytics Tools

The following section presents the Big Data Analytics tools and techniques adapted in this research work.

**Hadoop:**  Hadoop is an open source implementation of the Map Reduce framework that is commonly used by academic and industry for Big Data analysis. At the core of Hadoop are Hadoop Map Reduce and Hadoop Distributed File System (HDFS), the open source counterpart of Google File System (GFS). There are also a bundle of Hadoop-related projects supported by Apache Foundation, such as HBase (database), Hive (data warehouse), Pig (high- level data-flow), Zookeeper (high-performance coordination) and Mahout (scalable machine learning and data mining) [13]. Therefore, we choose Hadoop as the development platform to study the scalability of Naive Baye's classifier.

**Map Reduce:**  The Map Reduce structure has been used to process extensive datasets since the first paper [14, 15] was distributed. Google's clusters produce more than 20 Petabytes of information consistently by running one hundred thousand Map Reduce jobs on average [16]. Using this system, software engineers just need to concentrate on critical thinking. The Map Reduce runtime framework will deal with the underlining parallelization, adaptation to internal failure, information dissemination and load adjust. Google File System (GFS) is a distributed file system that Map Reduce uses for the storage of large data crosswise inexpensive hard drives. The accessibility and reliability of underlining questionable hardware are given by reproducing record pieces and distributing them crosswise over different hubs.

A Map Reduce work contains no less than a map function and a reduce function, called mapper and reducer. The mapper takes a couple of key/value as information and produces a set of key/value pairs. All key/value pairs are sorted by their keys and sent to various reducers as indicated by the key. Every reducer gets a key and a set of values that has the same key [15].

The Map Reduce is considered as a fantastic device for sorting or counting. The map and reduce functions are given the user a chance to implement their desired functionalities to process each key/value pair [15].

**Mahout:** Apache Mahout is one such intelligent application used to learn data from users. It helps to carry out machine learning operations in an effective and scalable manner. Mahout is thus a popular tool used to cluster documents, classify tags, organize content and perform collaborative filtering to give recommendations [17]. Therefore, we used Mahout to classify our tweeter dataset in this paper.

**R-Studio:** RStudio is a free and open-source integrated development environment (IDE) for R, a programming language for statistical computing and graphics. RStudio was founded by Allaire, [18] creator of the programming language ColdFusion. Hadley Wickham is the Chief Scientist at R Studio. We used R-studio in our research for fetching and processing the tweeter datasets to be classified later using Hadoop.

## 3    Methodology

The main aim of this paper is to help executives and higher management to make commercial plan and decisions based on the feedbacks of tweets from customer who is using their products. The goal is to classify streaming tweets to know customer opinion whether he is satisfied or not. For example, if a customer tweet is: "I am using IPhone 7 and I am pretty happy with it", this means that the customer is very happy and commercial plans for further services could be assigned to his accounts.

### 3.1    Proposed System for Classification of Tweets

To achieve the aim of effectively classifying tweets that contain information about people opinion about taking certain product, a recommender system is proposed, as (See Fig. 1).

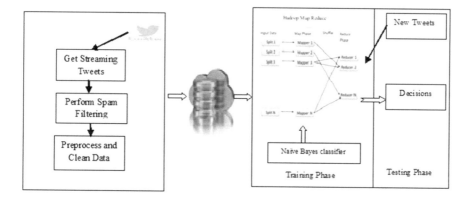

**Fig. 1.** Proposed recommender system for classifcation of tweets

The proposed system consists of two Modules, Preprocessing Module and Classification Module. Detail discussion is provided in the subsequent Sections.

## (1) Preprocessing Module

### (A) Extract Streaming Tweets about Certain Product

Fist we created an application account in Twitter that provides streaming API to allow us to retrieve real-time tweets. Then we used Twitter package in R-studio for accessing the Twitter API. We authorized R-studio to access the API of the precreated twitter application account. The product names need to be determined before retrieving tweets. We select product name as the keywords in tracking. In the selection process, we followed two rules. To have sufficient tweet about certain product, it should have been in the market for quite some time. The product should not be for customer that don't post tweets like children or people in amnesia [19]. As a case study, we selected one product with positive and negative feedback from customer, shown in Table 1. We used twitter package in r-studio to select only tweets of interest. For example, for the positive reviews, we fetched positive tweets by associating some positive words with it (example: "Good", "satisfied" "happy", etc.). Examples of fetched tweets are shown in Table 2.

**Table 1.** Information about retrieved tweets

| Product | Sentiments | Keywords | Fetched Tweets | Total | Language |
|---------|-----------|----------|----------------|-------|----------|
| IPhone | Positive | Happy | 2000 | 3000 | English |
| | | Satisfied | 1000 | | |
| | Negative | Bad | 638 | 2252 | |
| | | Ugly | 1000 | | |
| | | Unsatisfied | 614 | | |

**Table 2.** Examples of fetched tweets

| Sentiment | Example tweet | Key word |
|-----------|---------------|----------|
| Positive | RT @charmermark: Still very happy with my new iPhone SE. It rocks. | Happy |
| Negative | The red iPhone is so ugly | Ugly |

### (B) Perform Spam Filtering

After collecting the data, we performed an inspection test of the tweets. We commanded the R-studio to list 50 tweets of each review to inspect the irrelevant data that needs to be removed and cleansed. For example: we found in case ugly keywords, there was a lot of tweets that were meaning this words for the cover of their I-phone and not actual product i.e. iPhone. For example, the tweet "my iPhone case is ugly" is not a review about the IPhone rather it is about the phone case. We applied a rule to filter out spam tweets from the streamed data. A simple rule-based method [19] was used to this purpose, since spam filtering was not the main focus of this research project.

### (C) Pre-process and Clean the Data

Twitter tweets usually contain Uniform Resource Locators (URLs), emotions, hash tags, Internet slang, misspelled words and incorrect grammar expressions. R-Studio code was used to process the text. The following measures were taken to clean the tweets [19]:

- Hash tags and usernames were replaced with a space.
- Then all spaces were removed
- Punctuation characters were removed
- English stop words were removed
- Emojis were removed

### (D) Output the Tweets Dataset

The result of step (c) provides us a preprocessed cleansed data set, which can be used as an input to the classifier. Table 3 shows the cleaned finished dataset after removing hash tags, punctuations, English stop words, spaces and emojis.

**Table 3.** Total number of processed tweets

| Sentiment | Number of tweets |
|-----------|------------------|
| Positive | 2580 |
| Negative | 1015 |
| Total = 3595 | |

## (2) Classification Module

### (A) Training Phase

The output dataset from the firs module (Preprocessing Module) is sent as input to the second module (Classification Module) to be classified for decision making. The dataset will be considered as a training data set for the classifier. It is considered supervised learning because the labels of the classes are predefined as negative or positive reviews. In our proposed research, we have used Naïve Bayes classifier for classification of positive and negative tweets.

The Naïve Bayes machine learning technique using the Hadoop Mahout tool consists of the following process.

- **Step 1:** Convert the tweets to a sequence file so that it forms a <key,value> pair by creating sequence file using subdirectory utility.
- **Step 2:** Convert Sequence File to vectors by preprocessing the dataset into a <Text, VectorWritable> Sequence File containing term frequencies for each tweets document.
- **Step 3:** We split tweets dataset into two datasets, one for testing and one for training. Randomly splitting them for training 70% of records and for testing 30%
- **Step 4:** Training the classifier with the training set using Naïve Bayes algorithm.
- **Step 5:** Test the classifier against test dataset.

Table 4 shows sequence of steps performed on the Mahout tool.

**Table 4.** Syntax sample used for mahout Naive Bayes

| Step | Syntax |
|------|--------|
| Convert tweets data Sequence File | mahoutseqdirectory -i tweets -o tweets-input-seq |
| Convert Sequence File to vectors | mahout seq2sparse -i tweets-input-seq -o tweets-vectors |
| Split the data | mahout split -i tweets-vectors/tfidf-vectors --trainingOutput train-vectors -testOutput test-vectors --randomSelectionPct 30 --overwrite --sequenceFiles -xm sequential |
| Train the classifier | mahouttrainnb -i train-vectors -el -li labelindex -o model -owñc |
| Test classifier | mahouttestnb -i train-vectors -m model -l labelindex -ow -o tweet-testing -c |

The results of the trained and tested classifier will be the data model that will be used for classifying any new streaming tweets to decide whether a tweet contains a positive or negative review and helps in performing statistics over the model to support business decision making.

### (B) Testing Phase for new tweets

New tweets will be fed as input to the model. The results of the classifier will be the useful information for decision makers. These data can be further visualized and presented in smart life dashboards to support managers in creating specific or more attractive and valuable marketing plans.

## 4    Results

The aim of this paper is to provide a recommender system that will help decision makers in classifying tweets as it is highly significant for their marketing decisions plans. The model in general takes streaming tweets about certain product and then classifies it based on the predefined labels of positive or negative reviews. The results of this model, is a classifier data model that will be used in classifying any new tweets. The result generated from the model is shown in Fig. 2.

The data model has correctly classified 3573 tweets out of 3595 total tweets. The incorrectly classified instances were 22 tweets only, which is considered a small number. To specify more, the correctly classified instances were 2567 correctly identified positive tweets and 1006 correctly identified negative tweets. Moreover, the incorrectly classified tweets were 13 as positive tweets and 9 as negative tweets. The accuracy of the classifier is 99.39% which is a very high number that makes the classifier very successful as a target for identifying and classifying negative and positive reviews. The whole classifying process took 29286 ms (Minutes: 0.4881) which is considered very little processing time compared to the huge amount of data. Furthermore, the model can be used as a means for visualizing and presenting the data in a more understandable and appealing dashboard for decision makers.

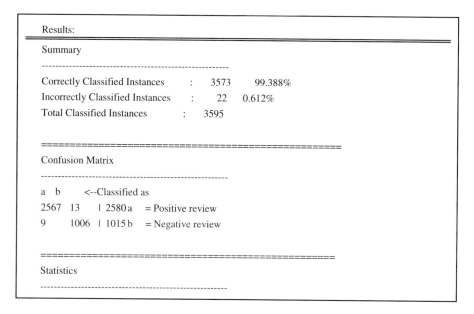

**Fig. 2.** The results generated from the classifier data model

## 4.1 Performance of the Classifier

The performance of the classifier is measured using Accuracy, Sensitivity and Specificity as shown in Eqs. (2), (3) and (4).

**Accuracy:** The accuracy of a test is its ability to differentiate the customer and tweets correctly. To estimate the accuracy of a test, we should calculate the proportion of true positive and true negative in all evaluated cases. Mathematically, this can be stated as shown in Eq. (2)

$$Accuracy = \frac{TP + TN}{TP + TN + FP + FN} \tag{2}$$

Where,

True positive (TP) = the number of tweets correctly identified as positive reviews.
False positive (FP) = the number of tweets incorrectly identified as positive reviews
True negative (TN) = the number of tweets correctly identified as negative reviews
False negative (FN) = the number of tweets incorrectly identified as negative reviews

In this case, the Hadoop Mahout program calculates the accuracy automatically, but we need to calculate the sensitivity and specificity as follows:

**Sensitivity:** The sensitivity of a test is its ability to determine the positive tweets correctly. To estimate it, we should calculate the proportion of true positive tweets. Mathematically, this can be stated as shown in Eq. (3).

$$\text{Sensitivity} = \frac{\text{TP}}{\text{TP} + \text{FN}} \tag{3}$$

The sensitivity in this case will be:

$$\text{Sensitivity} = 2567/(2567 + 9) = \mathbf{0.996}$$

**Specificity:** The specificity of a test is its ability to determine the negative tweets correctly. To estimate it, we should calculate the proportion of true negative in tweets. Mathematically, this can be stated as shown in Eq. (4).

$$\text{Sensitivity} = \frac{\text{TN}}{\text{TN} + \text{FP}} \tag{4}$$

The specificity in this case would be:

$$\text{Specificity} = 1015/(1015 + 13) = \mathbf{0.987}$$

See Figs. 3 and 4, which shows a pie graph for the percentage of correctly and incorrectly classified positive and negative tweets.

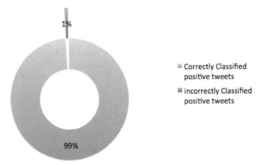

**Fig. 3.** Percentage of positive tweets

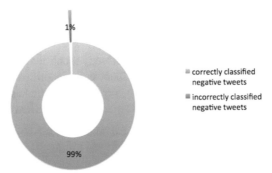

**Fig. 4.** Percentage of negative tweets

## 5    Conclusion

Twitter offers a promising median for marketing decision makers to plan their commercial campaign based on customer opinion. The nature of the twitter site in term of number of data per seconds and the lively streaming tweets require big data processor to deal with the analysis process. We proposed a Recommender system that provides information about the products that are likely to be helpful to the business decision makes, which performs the analysis within seconds. The model in general fetches the customer streaming tweets then classifies it based on the pre-defined labels of positive or negative review about certain product. High accuracy of the resultant classifier offers a great deal in correctly identifying any new tweets within minimal time.

## References

1. Data Analytics and Sentiment Analysis (2016). http://conferences.unicom.co.uk/data-and-sentiment-analysis/
2. Das, T.K., Acharjya, D.P., Patra, M.R.: Opinion mining about a product by analyzing public tweets in Twitter, pp. 3–6 (2014)
3. Vishal, A., Sonawane, P.: Sentiment analysis of twitter data: a survey of techniques. Int. J. Comput. Appl. (0975–8887) **139**(11), 5–15 (2016)
4. Pak, A., Paroubek, P.: Twitter as a corpus for sentiment analysis and opinion mining. In: Proceedings of the Seventh Conference on International Language Resources and Evaluation, pp. 1320–1326 (2010)
5. Agarwal, A., Xie, B., Vovsha, I., Rambow, O., Passonneau, R.: Sentiment analysis of twitter data. In: Proceedings of the ACL 2011 Workshop on Languages in Social Media, pp. 30–38 (2011)
6. Neethu, M.S., Rajashree, R.: Sentiment analysis in twitter using machine learning techniques. In: 4th ICCCNT, Tiruchengode, India. IEEE (2013)
7. Gautam, G., Yadav, D.: Sentiment analysis of twitter data using machine learning approaches and semantic analysis. In: Seventh International Conference on Contemporary Computing (IC3), India. IEEE (2014)
8. Amolik, A., Jivane, N., Bhandari, M., Venkatesan, M.: Twitter sentiment analysis of movie reviews using machine learning techniques. Int. J. Eng. Technol. (IJET) **7**, 1–7 (2016)
9. Go, A., Bhayani, R., Huang, L.: Twitter sentiment classification using distant supervision processing, pp. 1–6 (2009)
10. Manning, C.D., Schutze, H.: Foundations of Statistical Natural Language Processing. MIT Press, Cambridge (1999)
11. Geser, H.: Tweeted thoughts and twittered relationships. In: Sociology in Switzerland: Toward Cybersociety and Vireal Social Relations, Zuerich, February 2009
12. https://developer.twitter.com/en/docs/basics/things-every-developer-should-know
13. http://hadoop.apache.org
14. Blasch, E., et al.: Scalable sentiment classification for Big data analysis using Naïve Bayes classifier (2013)
15. Dean, J., Ghemawat, S.: Mapreduce: simplified data processing on large clusters. In: Proceedings of the 6th Conference on Symposium on Operating Systems Design & Implementation, OSDI 2004, Berkeley, CA, USA, vol. 6. USENIX Association (2004)

16. Dean, J., Ghemawat, S.: Mapreduce: simplified data processing on large clusters. Commun. ACM **51**(1), 107–113 (2008)
17. Apache Mahout: scalable machine learning and data mining. http://mahout.apache.org/
18. The R Project for Statistical Computing. https://www.rstudio.com/

# Evolution of Hashtags on Twitter: A Case Study from Events Groups

Layal Abu Daher[1(✉)], Rached Zantout[2], Islam Elkabani[1],
and Khaled Almustafa[3]

[1] Department of Mathematics and Computer Science, Faculty of Science,
Beirut Arab University, Beirut, Lebanon
{layalad,islam.kabani}@bau.edu.lb
[2] Electrical and Computer Engineering Department, Rafik Hariri University,
Mechref, Lebanon
zantoutrn@rhu.edu.lb
[3] College of Engineering, Prince Sultan University,
Riyadh, Kingdom of Saudi Arabia
kalmustafa@psu.edu.sa

**Abstract.** Twitter is a micro-blogging interactive platform that captured fame due to its simple features that allowed the communication between users. As the interactive communication grew bigger and faster, a feature called Hashtags became very known, famous and recognized by most of users as it acquainted the themes behind the posted tweets. This study focuses on the factors causing people to participate on some trendy hashtags on Twitter online social network. Consequently, these factors affect the evolution of such trendy hashtags over time. In order to study this evolution, a dataset reflecting real tweets from common events occurring between December 2015 and January 2016 were crawled. The reciprocal effect of users' topological features and activity levels has been studied. Two Influence Measures and one Topological Measure have been introduced in this work. Moreover, other measures available in the literature such as Activity Measures and Centrality Measures have been used. These measures, along with the three newly introduced measures contributed in the determination of the measures that might be influential for a user to attract other users to a certain hashtag. In this work, the focus is on the centrality levels in addition to the activity levels of users participating on the hashtags under study and the effect of those levels on the activity or the membership of other users on same hashtags.

**Keywords:** Evolution · Hashtags · Influence

## 1 Introduction

Nowadays, Online Social Networks are considered a new means of information propagation and sharing. Users on social media platforms share thoughts, life experiences, ideas and even needs on such social networks. Social network analysis is a method of analysis for social network graphs. From a graph topology perspective, the social network consists of a group of nodes (users) connected by edges (relationships).

© Springer International Publishing AG, part of Springer Nature 2018
M. Alenezi and B. Qureshi (Eds.): *5th International Symposium*
*on Data Mining Applications*, pp. 181–194, 2018.
https://doi.org/10.1007/978-3-319-78753-4_14

The relationship between nodes is either directed or undirected depending on the type of the social network graph. For example, Twitter, the most commonly used and investigated online social network [1], is a directed social network graph where the relationship between users does not have to be reciprocal. In other words, an incoming directed edge to a node from another node shows that the latter is following the first. According to a study done in [2], only 22% of connections between users on Twitter are reciprocal. On Twitter, users can initialize hashtags within their tweets by using a collection of characters preceded by a "#" symbol. This hashtag is a metadata tag that allows the tweets containing the text following the "#" to be grouped together by a specific subject such as political, commercial, entertaining or even events subjects. For instance, #Christmas was one of the most trendy hashtags used by users in the period between December 2015 and January 2016. Users initiated uncounted number of hashtags and posted uncounted number of tweets on hashtags about Christmas and other related events. Some of these initiated hashtags conserved their existence through users' postings, while others faded by time as users stopped posting on them. Our analytical study is conducted on a dataset collected from Twitter social network graph where some trendy hashtags about events occurring in the abovementioned period were collected in order to study the factors affecting their evolution.

The paper is organized as follows: in Sect. 2, related work is discussed; in Sect. 3, the approach is discussed focusing on data selection and representation in addition to measures identification and ranking; in Sect. 4, the results are presented and discussed and finally, conclusions and future work are presented in Sect. 5.

## 2 Related Work

Online Social Networks (OSNs) have developed to become a large part of our society. Browsing these networks became one of the daily tasks that people do on regular basis. This rapid growth of OSNs boosted researchers to explore and analyze the social network from all its aspects regarding evolution, prediction of users' participation, evaluation of influential nodes and measuring the influence of users on those OSNs.

A study done by [3] predicted the participation of users in social media. In this study, the main objective was to know if a topic will be interesting for users or not. They built association rules to predict the participation on a public group on Facebook. In addition, they used the association rule learning to discover the relationships between the users who are posting through commenting or liking on posts on Facebook groups. The results of this study showed high level of accuracy in predicting user participation on Facebook groups.

An interesting study done by [4] in 2016 examined the changes exhibited on the graph and the behavior of users since the first comprehensive study of Twitter done by [2] in the year 2010. This study revealed that Twitter popularity and usage has increased by 10-fold. The results of their study showed an increase in reciprocal edges between nodes revealing that 12.5% of 2009 Twitter users have left Twitter by the time the study was conducted. The authors also noticed the decrease of network connectivity and the movement of edges to popular users on the network. In addition, severe changes to influential users on Twitter has been noticed where non popular users in the past are considered the most popular and influential users by the time this study was conducted.

Another study by [5] measured user influence in Twitter social network by comparing these measures of influence which are in-degree, retweets, and mentions. This study investigated the user influence dynamics across topics and time. The results of this study revealed that most influential users hold significant influence over a variety of topics on Twitter. This means that influential users in the network can be influential if employed in different topics. In addition, the ordinary Twitter users can be influential if they focus on a single topic using smart, insightful and creative posts that are spotted as valuable posts to other users on the network. The findings of this study were very effective in the marketing, the political and emotional sectors where influential users might be of added value to the sector they are active on, and ordinary users with even single post might be influential enough to convince other users with a certain point of view.

A study done by [6] focusing on content based prediction of the spread of ideas in microblogging communities. In this work, the authors captured the acceptance of a hashtag by the count of its appearance in a time interval and as a result, they predicted its frequency after some time. They used a data set of 4 million tweets and thousands of hashtags. The results of this study show that there are three main factors to accept an idea in a hashtag: The content of an idea, the context within an idea and the social graph. What makes this work different is that it is the first study that focuses on the content features in addition to the framework presented which efficiently modeled the exposure and acceptance of an idea in Twitter hashtags.

The study done in [7] revealed that analyzing social networks leads to understanding the structure and dynamics of social groups. In addition, the process by which people come together, call for new members, and change over time is a central subject of study in social science especially in politics and religion. Thus, group evolution and predicting influential users were popular fields of research since the last decade.

For instance, in [8], a new measure of group proximity in OSNs was introduced where the link cardinality between groups was estimated. Therefore, by using friendship links and mutual members between groups, the evolution of such groups could be estimated.

Another work done by [9] predicted the stability of groups in the dataset collected from "World of Warcraft" multiplayer online game. Their aim was to predict if a group on the dataset collected will stay the same or shrink over time.

Another study [10] in the area of group evolution conducted experimental studies on four evolving social networks in which ten different classifiers were used. They used the GED method used in [11] to discover group evolution where inclusion, as a measure, was the most important component. This measure allowed evaluating the inclusion of one group in another depending on the group quantity (what portion of members from the first group is in the second group) and quality (what contribution of important members from the first group is in the second group). Such a measure shows a balance between the groups containing many of less important members and the groups of only few but key members. The authors could extract the sequence of group sizes and events between time frames through following the GED method steps.

The study in [12] investigated the disappearance of a group of nodes in a social network. This was important to avoid the loss of information that could follow the disappearance of a node and to identify a substitute to its disappearance whenever it is critical in terms of its centrality measures. In other words, a potential substitute of the disappearing node along with the new links is found between individuals that share common characteristics such as values, education and beliefs. Such an approach adds enough links to maintain the quality of the information flow within the network as it was before a node deletion. The consequences of the steps are: success and maintaining the network connected after the disappearance of a set of nodes, reasonable execution times after a node deletion, and constant quality of the information flow.

Focusing on the prediction of community evolution, in [13], the process of group prediction was analyzed by conducting experimental studies on three different datasets: Facebook, DBLP and Salon24-Polish blogosphere. Four main bases were covered in the study: the first phase included data collection and splitting data into time frames. The second phase consisted of classifying the social networks for each period and identifying the social community. The third phase concentrated on identifying the changes or events in a group and detecting the chains preceding the recent state of the group. The last phase was characterized by building the predictive model. This was done by learning the classifier and validating it. In addition, two algorithms were used: Stable Group Changes Identification (SGCI) and Group Evolution Discovery (GED). The experimental studies on the three different databases revealed that as the length of the evolution chain increases, the accuracy of prediction is improved. This means that, more comprehensive history with more information about the group is detected by longer evolution chains. The best results for SGCI method using either Random Forest or AdaBoost classifier were recorded for Salon24 dataset with evolution chains of length 6 and longer. As to GED method, the best predictive ability was also recognized for Salon24 dataset with evolution chains of length 7 and using the same classifiers used in SGCI method.

Various studies were conducted on different OSNs in order to predict the influential users. Some of the studies focused on comment mining as in [14] and others concentrated on predicting information cascade as in [15]. Identifying influential users on OSNs is a very important field in social network analysis. In the first place, users are the most important factors in social networks, hence pointing out influential among these users can help in the prediction of the evolution of such networks because these users might be a strong clue for the future of such groups.

Centrality measures determine how and to whom this user is connected. Closeness and Betweenness centrality measures are the most popular centrality measures used in the literature. *Closeness* is the minimum sum of the shortest paths from a node to all other nodes in a network [16]. *Betweenness* is the number of shortest paths passing through a vertex [17].

# 3   Approach

The dataset presented in this work constitutes of posts on specified hashtags from Twitter and downloaded using the crawler designed in [18]. Basic information were downloaded about the posting users such as: the original tweet text, the retweet text, and the number of followers of the posting users. In order to collect the followers of the posting users, we built our own crawler. We used "hashtagify.me" website [19] to choose our dataset of hashtags. This website identifies the most trendy hashtags within the past month and presents the percentage of correlation between these trendy hashtags and other 10 hashtags. Accordingly, we extracted "#Christmas", one of the top trendy hashtags during the period covering December 2015 and January 2016, in addition to the ten correlated hashtags as displayed in Table 1.

**Table 1.** The 10 correlated #s to #Christmas with % of correlation

| Hashtags | Correlation |
|---|---|
| #xmas | 5.6% |
| #win | 4.5% |
| #gift | 4.0% |
| #giveaway | 2.4% |
| #Santa | 2.3% |
| #gifts | 2.1% |
| #competition | 1.9% |
| #holiday | 1.7% |
| #love | 1.6% |
| #handmade | 1.6% |

## 3.1 Data Selection

In the hashtags collected for analysis, there exist 58361 unique users with 34702 tweets and 47818 retweets as shown in Table 2. A percentage of 29.79% of posting users on these hashtags had private profiles (about 17389 out of 58361) where they were ignored in the preprocessing phase since crawling their basic information failed.

In order to build the corpus of our analytical study, we selected the top hashtags that have the highest percentage of correlation with the root hashtag "#Christmas"

**Table 2.** Number of Tweets and Retweets in the 11 #s under study

| Hashtags | Tweets | Retweets |
|---|---|---|
| #Christmas | 7089 | 6280 |
| #xmas | 5280 | 6469 |
| #win | 3528 | 16138 |
| #gift | 4347 | 2148 |
| #giveaway | 4018 | 6917 |
| #Santa | 1892 | 1353 |
| #gifts | 1719 | 1678 |
| #competition | 362 | 2324 |
| #holiday | 2141 | 1587 |
| #love | 3078 | 1780 |
| #handmade | 1248 | 1144 |

along with the hashtags with highest number of tweets and retweets. Our corpus resulted in three hashtags: #Christmas, #xmas and #win. The other remaining eight hashtags were removed from our analytical study due to the low percentage of correlation and the inadequate number of tweets and retweets as shown in Table 2. The three chosen hashtags are also the most general ones, where the data collected varies in terms of the number of posting users, tweets, retweets and followers.

## 3.2   Data Representation

To represent and visualize the collected dataset, Gephi software is used [20]. This software represents dataset through real time analysis and visualization by creating links between different posting users on the hashtags collected. The directed graph is represented by source nodes, target nodes and weight of the edge between both nodes. In our study, we consider that there is a directed edge from user A to user B if user A is found in the list of followers of user B. As to the weight of edges between nodes, we considered the Jaccard's similarity coefficient presented in [21] as the weight of edge. The Jaccard's Similarity between two nodes A and B is calculated using Eq. 1.

$$J(A, B) = \frac{followers(A) \cap followers(B)}{followers(A) \cup followers(B)} \tag{1}$$

## 3.3   Identifying Measures

In order to study the social network graph, two *Centrality Measures* are used from the literature namely, *Betweenness* and *Closeness Centrality*. Moreover, three *Activity Measures* are also used. Two new *Influence Measures* are introduced in this paper namely, *Count of Posting Followers (CPF)* and *Users Being Retweeted (UBR)*. Another new *Topological Measure* is newly introduced in this work namely, *Jaccard Weighted In-degree Measure*.

**Influence Measures**
The two new *Influence Measures* introduced in this work are *Count of Posting Followers (CPF)* and *Users Being Retweeted (UBR)*. CPF finds the total number of followers of user A who are posting on the same hashtag where user A is posting as shown in Eq. 2.

$$CPF = \sum_{i \varepsilon F(A)} x_i \tag{2}$$

where F(A) is the set of followers of User A.

The second *Influence Measure*, namely *UBR* finds the sum of the Retweets of the tweets of user A as represented in Eq. 3.

$$UBR(A) = \sum_{t \varepsilon T(A)} RT(t) \tag{3}$$

where T(A) denotes the set of tweets by user A

RT(t) denotes the number of retweets of a tweet t

## Activity Measures

The activity of users on Twitter online social network is calculated by summing his real participation through his tweets and retweets. Many twitter users are only readers where their activity will not be counted on the network without real participation of posting or sharing.

The simplest *Activity Measure* of a user A is his number of tweets on a certain hashtag. Another simple *Activity Measure* is the number of retweets of user A. Thus, the *Activity Measure* is the sum of all the actions of a user on a certain hashtag and is referred to as the *General Activity* as stated in [22] and illustrated in Eq. 4.

$$General\ Activity(A) = T + RT \tag{4}$$

where T is the number of tweets of User A

RT is the number of Retweets of user A

Another *Activity Measure* introduced in [23] is *Topical Signal (TS)*. *TS* is the ration of the *General Activity* of a certain active user by the total number of tweets in a specific hashtag as shown in Eq. 5.

$$TS(A) = \frac{(T + RT)_{|specific\ Hashtags}}{Total\ Numer\ of\ Tweets} \tag{5}$$

A third *Activity Measure* is used in this paper and was also introduced in [23] which is *Signal Strength (SS)*. *SS* determines how strong is the *Topical Signal* as shown in Eq. 6. It is calculated as the ratio of the original tweets to the summation of original tweets and retweets of a user A.

$$SS(A) = \frac{T}{T + RT} \tag{6}$$

## Topological Measures

Two *Topological Measures* are used in this paper, the *Centrality Measures* and *Jaccard's weighted In-degree* measure *(JWI)*. *JWI* is a new measure introduced in this paper as a better measure for user social connectedness. It is calculated by summing the Jaccard's similarity weights of all incoming edges to user A, as shown in Eq. (7).

$$JWI(A) = \sum_{i=1}^{n} w_i(A) \tag{7}$$

where *w* is the weight of the edge between two nodes.

In our study, we use the *Betweenness* and *Closeness Centrality* measures to study the relationship between the centrality of the user in a social network and his influence on the evolution of a certain hashtag.

### 3.4   Ranking Based on the Identified Measures

A directed edge between two users is created if one of the users is found in the list of followers of the second. For this graph, the *Influence Measures* (*CPF & UBR*), *Activity Measures* (*GA, TS, SS*) and *Topological Measures* (*Centrality measures* and *JWI*) are calculated. Ordered (descending) lists are created and the top ranked users identified from *CPF Measure* and the top ranked users identified from *Activity Measures* are compared. Similarly, the same comparison is done between the top ranked users identified from the *CPF Measure* and the top users identified from the *Topological Measures*. Also, comparison is done between the top ranked users identified from the *UBR Measure* and the top users identified from the *Activity Measures*. In addition, comparison is also done between the top ranked users identified from the *UBR Measure* and the top users identified from the *Topological Measures*.

## 4   Results and Discussion

In this analytical study, we emphasize on studying relationships between the levels of *Activity Measures*, *Centrality Measures* and *JWI Measure* on one hand and the levels of the two measures, *CPF* and *UBR*. This section will demonstrate the results of intersection between the top users ranked by *Activity Measures* and the top users ranked by *CPF* and *UBR* for different percentages of top users (25%, 50% and 75%). Analysis focused on the 25% of top users with conclusions reinforced by the other percentages of top users.

### 4.1   Results

Table 3 shows the results of the *CPF* intersection with the level of the three *Activity Measures* namely *General Activity*, *Topical Signal* and *Signal Strength* for the three hashtags under study. The *CPF* intersection with the *Activity Measures* did not exceed 35% in the three hashtags under study as shown in the column labelled *25% intersection* in Table 3. This means that the followers of a user are not posting on the same hashtag because of the influence of other user's activity. Comparing the intersection results across different percentages of top users (i.e. 25%, 50%, and 75%), a logical observation can be made that it increases linearly.

The results of the *CPF* intersection with the level of the three *Topological* Measures, namely *Betweenness Centrality*, *Closeness Centrality* and *JWI* for the three selected hashtags are shown in Table 4. The *CPF* intersection with the *Betweenness Centrality* and the *Closeness Centrality Measures* reached 80% and reached 74% in *JWI* in #Christmas. As to the intersection results for #Xmas, *CPF* intersection with the *Betweenness* and the *Closeness Centrality Measures* was between 64% and 77% and the *JWI* reached 71%. The *CPF* intersection with the *three Topological Measures* in

**Table 3.** Similarity between *CPF* and Activity Measures

*CPF* Intersection Results

*#Christmas Results*

| Activity Measures | 25% ∩ (2713 Users) | | 50% ∩ (5426 Users) | | 75% ∩ (8138 Users) | |
|---|---|---|---|---|---|---|
| | No. | % | No. | % | No. | % |
| General Activity | 786 | 29 | 3072 | 57 | 6151 | 76 |
| Topical Signal | 786 | 29 | 3072 | 57 | 6151 | 76 |
| Signal Strength | 743 | 27 | 2775 | 51 | 6797 | 84 |

*#Xmas Results*

| Activity Measures | 25% ∩ (2264 Users) | | 50% ∩ (4528 Users) | | 75% ∩ (6792 Users) | |
|---|---|---|---|---|---|---|
| | No. | % | No. | % | No. | % |
| General Activity | 799 | 35 | 2466 | 54 | 5740 | 85 |
| Topical Signal | 799 | 35 | 2466 | 54 | 5740 | 85 |
| Signal Strength | 514 | 23 | 2183 | 48 | 5678 | 84 |

*#Win Results*

| Activity Measures | 25% ∩ (1268 Users) | | 50% ∩ (2537 Users) | | 75% ∩ (3805 Users) | |
|---|---|---|---|---|---|---|
| | No. | % | No. | % | No. | % |
| General Activity | 368 | 29 | 1365 | 54 | 2919 | 77 |
| Topical Signal | 368 | 29 | 1365 | 54 | 2919 | 77 |
| Signal Strength | 438 | 35 | 1461 | 58 | 2997 | 79 |

#Win reached 77% and the *JWI* measure reached 72%. This indicates that the more socially connected users are more likely to affect their followers to post on the same hashtags. Comparing the intersection results across different percentages of top users (i.e. 25%, 50%, and 75%), a logical observation can be made that it increases linearly in most of the cases.

Table 5 shows the results of the *UBR* intersection with the level of the three *Activity Measures*, namely *General Activity*, *Topical Signal* and *Signal Strength* for the three selected hashtags. In #Christmas and #Win, the three activity measures reached 40%. However, in #Xmas, *General Activity* and *Topical Signal* indicated 77% whereas the *Signal Strength* reached 47%. These results indicate that in #Xmas, the users' original tweets have more influence than the users' activity. Comparing the intersection results across different percentages of top users (i.e. 25%, 50%, and 75%), a logical observation can be made that it increases linearly in most of the cases.

Table 6 presents the results of the *UBR* intersection with the level of the three *Topological Measures*, namely *Betweenness Centrality*, *Closeness Centrality* and *JWI* for the three selected hashtags. The *UBR* intersection with *Topological Measures* did not exceed 36% in #Christmas, #Xmas and #Win. This indicates that the users who are

**Table 4.** Similarity between *CPF* and *Topological Measures*

*CPF* Intersection Results

*#Christmas Results*

| Topological Measures | 25% ∩ (2713 Users) | | 50% ∩ (5426 Users) | | 75% ∩ (8138 Users) | |
|---|---|---|---|---|---|---|
| | No. | % | No. | % | No. | % |
| Betweenness Centrality | 2177 | 80 | 5208 | 96 | 8069 | 99 |
| Closeness Centrality | 2166 | 80 | 5208 | 96 | 8069 | 99 |
| JWI | 2019 | 74 | 5028 | 93 | 8069 | 99 |

*#Xmas Results*

| Topological Measures | 25% ∩ (2264 Users) | | 50% ∩ (4528 Users) | | 75% ∩ (6792 Users) | |
|---|---|---|---|---|---|---|
| | No. | % | No. | % | No. | % |
| Betweenness Centrality | 1754 | 77 | 4526 | 100 | 6571 | 97 |
| Closeness Centrality | 1442 | 64 | 4526 | 100 | 6571 | 97 |
| JWI | 1614 | 71 | 3772 | 100 | 6424 | 95 |

*#Win Results*

| Topological Measures | 25% ∩ (1268 Users) | | 50% ∩ (2537 Users) | | 75% ∩ (3805 Users) | |
|---|---|---|---|---|---|---|
| | No. | % | No. | % | No. | % |
| Betweenness Centrality | 913 | 72 | 2291 | 90 | 3515 | 92 |
| Closeness Centrality | 973 | 77 | 2244 | 88 | 3804 | 100 |
| JWI | 918 | 72 | 2169 | 85 | 3533 | 93 |

more retweeted are not necessarily the ones who are socially connected. Comparing the intersection results across different percentages of top users (i.e. 25%, 50%, and 75%), a logical observation can be made that it increases linearly in most of the cases.

### 4.2    Discussion

Interesting findings were revealed from the results described in Tables 3, 4, 5 and 6. Starting by the influence of activity levels, the followers of the users posting on the same hashtag are not necessarily due to the activity levels of those users. Moreover, the *JWI* measure along with the two *Centrality Measures* used in this study were effective measures and influential to cause the followers of a user to post on the same hashtags. This means that as the connectivity between users increases, the influence of users on their followers also increases. Although the three *Topological Measures* proved to be a better indicator for the influence of users on their followers to post on same hashtags, these *Topological Measures* were not a good indicator for identifying users being retweeted. This is obvious from the results in Tables 4 and 6 where the most socially connected users were not retweeted but were among the users having the most posting

**Table 5.** Similarity between *UBR* and Activity Measures

| *UBR* Intersection Results | | | | | | |
|---|---|---|---|---|---|---|
| **#Christmas Results** | | | | | | |
| Activity Measures | 25% ∩ (2713 Users) | | 50% ∩ (5426 Users) | | 75% ∩ (8138 Users) | |
| | No. | % | No. | % | No. | % |
| General Activity | 945 | 35 | 4046 | 75 | 6303 | 77 |
| Topical Signal | 945 | 35 | 4046 | 75 | 6303 | 77 |
| Signal Strength | 961 | 35 | 2946 | 54 | 6869 | 84 |
| **#Xmas Results** | | | | | | |
| Activity Measures | 25% ∩ (2264 Users) | | 50% ∩ (4528 Users) | | 75% ∩ (6792 Users) | |
| | No. | % | No. | % | No. | % |
| General Activity | 1735 | 77 | 4210 | 93 | 6591 | 97 |
| Topical Signal | 1735 | 77 | 4210 | 93 | 6591 | 97 |
| Signal Strength | 1074 | 47 | 2929 | 65 | 5717 | 84 |
| **#win Results** | | | | | | |
| Activity Measures | 25% ∩ (1268 Users) | | 50% ∩ 2537 Users) | | 75% ∩ (3805 Users) | |
| | No. | % | No. | % | No. | % |
| General Activity | 511 | 40 | 1493 | 59 | 3276 | 86 |
| Topical Signal | 511 | 40 | 1493 | 59 | 3276 | 86 |
| Signal Strength | 423 | 33 | 1882 | 74 | 3466 | 91 |

followers. As to the results of intersection between *Activity Measures* and *UBR* for the three hashtags, we noticed that in #Xmas, as the activity of users posting on this hashtag increases, they are being retweeted more; however, they are not affecting their followers to post. This indicates that in #Xmas, retweeting is more influenced by the activity of users on this hashtag and by the total number of tweets available on this hashtag. Whereas, in #Christmas and #Win, the level of the three *Activity Measures* did not give a strong indicator for the users being retweeted. The vast majority of the intersections increased linearly with the percentage of top users. However, some outliers existed because there were many users having the same rank in the top 50% and 75%. This is why we focused on the top 25% in the analysis of the similarity comparison results.

**Table 6.** Similarity between *UBR* and *Topological Measures*

| UBR Intersection Results | | | | | | |
|---|---|---|---|---|---|---|
| **#Christmas Results** | | | | | | |
| Topological Measures | 25% ∩ (2713 Users) | | 50% ∩ (5426 Users) | | 75% ∩ (8138 Users) | |
| | No. | % | No. | % | No. | % |
| Betweenness Centrality | 921 | 34 | 3328 | 61 | 7654 | 94 |
| Closeness Centrality | 927 | 34 | 3328 | 61 | 7654 | 94 |
| JWI | 880 | 32 | 3321 | 61 | 7654 | 94 |
| **#Xmas Results** | | | | | | |
| Topological Measures | 25% ∩ (2264 Users) | | 50% ∩ (4528 Users) | | 75% ∩ (6792 Users) | |
| | No. | % | No. | % | No. | % |
| Betweenness Centrality | 780 | 34 | 2525 | 56 | 5630 | 83 |
| Closeness Centrality | 742 | 33 | 2525 | 56 | 5630 | 83 |
| JWI | 734 | 32 | 2663 | 59 | 5889 | 87 |
| **#Win Results** | | | | | | |
| Topological Measures | 25% ∩ (1268 Users) | | 50% ∩ (2537 Users) | | 75% ∩ (3805 Users) | |
| | No. | % | No. | % | No. | % |
| Betweenness Centrality | 458 | 36 | 1420 | 56 | 3119 | 82 |
| Closeness Centrality | 446 | 35 | 1401 | 55 | 2986 | 78 |
| JWI | 438 | 35 | 1408 | 55 | 2890 | 76 |

## 5   Conclusion and Future Work

The analytical work in this paper investigates the factors affecting the participation of users in a selection of Twitter hashtags. Two *Influence Measures* and one *Topological Measure* are newly introduced. Those measures were proven to be better at measuring the influence and activity of a user in a hashtag. Our corpus consisted of three hashtags in order to analyze the behavior of users in Twitter. We downloaded basic information about users posting on those hashtags in addition to their tweets and retweets. This study revealed interesting findings where the level of activity was found not to affect the count of posting followers. Furthermore, in #Xmas, retweeting was found to be more influenced by the activity of users and the total number of tweets than by the social connectedness between users. Third, socially connected users proved to be more influential causing their followers to post on the same hashtags.

Future work will concentrate on studying user's behavior in order to predict participation, moreover, identifying influential users using association rule learning will be tackled.

# References

1. Tufekci, Z.: Big questions for social media big data: representativeness, validity and other methodological pitfalls. In: Proceedings of the 8th International AAAI Conference on Weblogs and Social Media, ICWSM 2014 (2014)
2. Lee, C., Kwak, H., Park, H., Moon, S.: What is Twitter, a social network or a news media? In: Proceedings of the 19th International Conference on World Wide Web, WWW 2010, Raleigh, North Carolina, USA (2010)
3. Erlandsson, F., Borg, A., Johnson, H., Bródka, P.: Predicting user participation in social media. In: Proceedings of the 12th International Conference and School on Advances in Network Science, NetSci-X 2016, Wroclaw, Poland (2016)
4. Efstathiades, H., Antoniades, D., Pallis, G.: Online social network evolution: revisiting the Twitter graph. In: 2016 IEEE International Conference on Big Data (Big Data), Washington, DC (2016)
5. Cha, M., Haddadi, H., Benevenuto, F., Gummadi, K.P.: Measuring user influence in twitter: the million follower fallacy. In: 4th International AAAI Conference on Weblogs and Social Media, Washington, DC (2010)
6. Tsur, O., Rappoport, A.: What's in a hashtag? Content based prediction of the spread of ideas in microblogging communities. In: Proceedings of the Fifth ACM International Conference on Web Search and Data Mining, WSDM 2012, Seattle, Washington, USA (2012)
7. Backstrom, L., Huttenlocher, D., Kleinberg, J.: Group formation in large social networks: membership, growth, and evolution. In: Proceedings of the 12th ACM SIGKDD International Conference on Knowledge Discovery and Data Mining, KDD 2006, Philadelphia, PA, USA (2006)
8. Saha, B., Getoor, L.: Group proximity measure for recommending groups in online social networks. In: The 2nd SNA-KDD Workshop, SNA-KDD 2008, Las Vegas, Nevada, USA (2008)
9. Patil, A., Liu, J., Gao, J.: Predicting group stability in online social networks. In: Proceedings of the 22nd International Conference on World Wide Web, WWW 2013, Rio de Janeiro, Brazil (2013)
10. Bródka, P., Kazienko, P., Kołoszczyk, B.: Predicting group evolution in the social network. In: International Conference on Social Informatics, Berlin, Heidelberg (2012)
11. Bródka, P., Saganowski, S., Kazienko, P.: GED: the method for group evolution discovery in social networks. Soc. Netw. Anal. Min. **3**(1), 1–14 (2013)
12. Sarr, I., Missaoui, R., Lalande, R.: Group disappearance in social networks with communities. Soc. Netw. Anal. Min. **3**(3), 651–665 (2013)
13. Saganowski, S., Gliwa, B., Bródka, P., Zygmunt, A., Kazienko, P., Koźlak, J.: Predicting community evolution in social networks. Entropy **17**(5), 3053–3096 (2015)
14. Jamali, S., Rangwala, H.: Digging digg: comment mining, popularity prediction, and social network analysis. In: Proceedings of the 2009 International Conference on Web Information Systems and Mining, Washington, DC, USA (2009)
15. Hakim, M.A.N., Khodra, M.L.: Predicting information cascade on Twitter using support vector regression. In: Proceedings of the 2014 International Conference on Data and Software Engineering (ICODSE), Hyderabad, India (2014)
16. Yan, E., Ding, Y.: Applying centrality measures to impact analysis: a coauthorship network analysis. J. Am. Soc. Inf. Sci. Technol. **60**(1), 2107–2118 (2009)
17. Freeman, L.C.: Centrality in social networks: conceptual clarification. Soc. Netw. **1**(3), 215–239 (1979)

18. Hawksey, M.: The musing of Martin Hawksey (EdTech Explorer), WordPress and Stargazer (2011). https://mashe.hawksey.info/. Accessed 10 Nov 2015
19. Hashtagify: Find, Analyse, Amplify, CyBranding Ltd. (2011). https://hashtagify.me/explorer/. Accessed 1 Mar 2015
20. Bastian, M., Heymann, S., Jacomy, M.: Gephi: an open source software for exploring and manipulating network. In: Proceedings of the Third International Conference on Weblogs and Social Media, ICWSM 2009, San Jose, California, USA (2009)
21. Niwattanakul, S., Singthongchai, J., Naenudorn, E., Wanapu, S.: Using of Jaccard coefficient for keywords similarity. In: Proceedings of the International MultiConference of Engineers and Computer Scientists, Hong Kong, China (2013)
22. Riquelme, F., González-Cantergiani, P.: Measuring user influence on Twitter: a survey. Inf. Process. Manage. **52**(5), 949–975 (2016)
23. Pal, A., Counts, S.: Identifying topical authorities in microblogs. In: Proceedings of the Fourth ACM International Conference on Web Search and Data Mining, WSDM 2011, Hong Kong, China (2011)

# Bioinformatics (Bio)

# Investments in Deep Learning Techniques for Improving the Biometric System Accuracy

A. Meraoumia[1], S. Chitroub[2(✉)], O. Chergui[1,3], and H. Bendjenna[1]

[1] LAboratory of Mathematics, Informatics and Systems (LAMIS),
University of Larbi Tebessi, 12002 Tebessa, Algeria
ameraoumia@gmail.com, hbendjenna@gmail.com
[2] LISIC Laboratory, Telecommunication Department,
Electronics and Computer Science Faculty, USTHB,
El-Alia, BP. 32, 16111 Bab-Ezzouar, Algiers, Algeria
s_chitroub@hotmail.com
[3] Ecole nationale Superieure d'Informatique (ESI),
BP 68M, 16309 Oued Smar, El Harrach, Algeria
o_chergui@esi.dz

**Abstract.** Recently, user identification is an essential foundation for protecting information in several applications. However, the need for heightened this identification has expanded the research to focus on the biometric traits of the users. Traditionally, identity determination is assured by recognizing personals used classical security means such password and/or card. Recently, a natural and reliable solution to this problem is offered by biometrics technologies. So, among several biometric modalities, these extracted from the hand have been systematically used to make identification for last years. In other hand, all issues related to the final conception of a biometric system are generally related to the classification task (classifier used). In this paper, we present a Restricted Boltzmann Machine (RBM) for palmprint (PLP) and palm-vein (PLV) identification which are able to classify precisely these modalities. In addition, in order to improving the identification system accuracy, a RBM based Deep Belief Nets (DBN) is also presented. The obtained results, using databases of 400 persons, have showed that deep learning methods has higher performances compared to the classical methods developed in the literature in terms of systems accuracies.

**Keywords:** Biometrics · Security · Deep learning
Restricted Boltzmann machine · Deep belief networks · Data fusion

## 1 Introduction

The person's identities recognition is one of the means to ensure the security of the logical or physical access as well as information systems, which has become

© Springer International Publishing AG, part of Springer Nature 2018
M. Alenezi and B. Qureshi (Eds.): *5th International Symposium on Data Mining Applications*, pp. 197–209, 2018.
https://doi.org/10.1007/978-3-319-78753-4_15

more and more of a very important priority [1]. However, it is almost the important and effectively used way to ensure security. In fact, due to the great need for such recognition, several ways that are related to the information that a person has or knows are developed. The first one is based on password and PIN code (*knowledge-based techniques*). Whereas, the second is based on key and card (*token-based techniques*). A password or a PIN code may be forgotten, guessed or pirated; while a key or card may be stolen, lost or forged. Therefore, some users are able to falsify their identity and enter to a system, or a place in which they are not allowed [2]. Unlike the above mentioned security means, biometrics is based on characteristics related to the person himself. It constitutes a strong and permanent link between a person and his identity; it cannot be stolen, forgotten, lost or passed on to another person. Due this reason, it is considered the strongest way compared with the conventional means which are based on knowledge or token. Thus, each person has its own biometric characteristics and they are relatively stable hence the idea of using these unique and unique features for identity recognition [3]. Therefore, the purpose of biometrics is to automatically recognize or verify a person's identity in order to assure the access to information (logical or physical access). Using behavioral, biological or physical characteristics related to person, several modalities biometrics can be extracted and then used for discriminating him from other [4].

Indeed, majority of researches on biometrics systems based on PLP have focused on their images captured under visible light. However, during the past few years, some researchers have used more features from the palm, such as the veins of the palm to improve the effect of these systems. These additionally features, which are the inner vessel structures beneath the skin or simply veinnets, are extracted from multispectral representation of the PLP image using near-infrared light. Obviously, the palm-veins are a low risk of falsification, difficulty of duplicated and stability because they lie under the skin. Moreover, the availability of a device that can acquire PLP and PLV images simultaneously has promoted research to constrict a multimodal biometric system based on the fusion of PLP and PLV modalities in order to overcome some of the limitations imposed by unimodal biometric systems like as insufficient accuracy caused, for example, by noisy data acquisition in certain environments [5].

During the last years, several works in the literature point out the hard works to improve the biometric systems, resulting in many systems architectures which are used different algorithms. Nowadays, a number of artificial intelligence algorithms have been emerged and the machine learning is one of these algorithms. Thus, one of the powerful techniques in this field is the deep machine learning. Thus, deep learning has found application in a large range of pattern recognition works [6]. It is subfield of Machine Learning research which applied learning algorithms to scan multiple levels of representation to modeling complex relation within the data [7]. Also, it allows a computer model to learn how to perform classification tasks directly from images, text, or audio by providing it with examples of data on which it trains. Models are trained through a large set of labeled data and neural network architectures that contain many layers. Deep Learning models can achieve an exceptional level of accuracy. Logically, many researchers have attempted to use these new techniques in order to improve the biometric system performance. After several attempts, researchers have finally arrived at successful attempts such as DBN structure. In this paper, to show the discriminating capability of the deep learning technique motioned above, two biometric systems were proposed. However, the two systems are based on the hand biometrics modalities which are the PLP and the PLV. In our study, two techniques, which are RBM and DBN, are evaluated and then from the point of view of identification system accuracies. Our systems are tested and evaluated using a popularly PLP and PLV databases that contains 400 persons. So, the obtained results prove the effectiveness and reliability of the DBN as one of the most powerful technique which can used in several high security application.

The remainder of this paper is organized as follows: Sect. 2 describes system design and feature extraction process. Section 3 presents the experimental results in order to evaluate the performance of the proposed system in two identification modes when using two deep learning methods. Finally, conclusions and perspectives are given in the last section.

## 2    System Framework

Like all biometric systems, our proposed PLP and PLV based biometrics systems (Fig. 1 describes an example of PLP based system) includes several processes which are pre-processing process, feature extraction process, feature matching process and decision process. In fact, in any biometric system two main phases are performed [8]: enrollment and recognition. Thus, to enroll into the system database, users have to provide a set of training PLP and/or PLV images. Typically, feature vector is extracted from each image using an appropriate feature extraction technique. In the identification phase, the same feature vectors are extracted from the test PLP and/or PLV images and the degree of similarity is computed using all of feature vectors references in the system database. In order to prepare the palmprint image to the feature extraction phase, a preprocessing step is required. The role of this step is to extract only the area information of the modality, called Region Of Interest (ROI) sub-image [9].

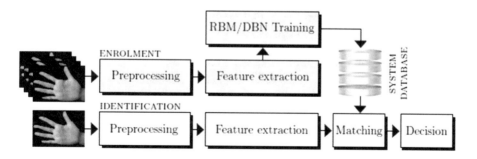

**Fig. 1.** Palmprint based biometric system using RBM and DBN modeling process.

## 3    Features Extraction and Modeling

Feature extraction is an important task in the pattern recognition applications due to the large amount of different existing features in signal (especially in image) and his multiple areas of applications that it covers medical imaging, biometrics, remote sensing and robotics. Due this necessity, a well considerable effort has been made by the researchers in this direction, resulting in many cases an excellent enough classification results. Furthermore, the extracted vector can be used directly by comparing it to another feature vector using a distance based feature matching process or as an input vector for an appropriate model.

### 3.1    Features Extraction Process

In our work, two modeling structures are used. Thus, the first one (RBM) use directly the image as an input vectors (arranged into one-dimensional vector), whereas, the second (DBN) use a binary image as an input. For that, it is necessary to use a feature extraction method which provides a binary vector. Thus, various features, including geometrical features, lines, singular points and texture, can be extracted and then used as a feature vector for discriminating the image pattern. So, methods based on lines structure have achieved promising results and high accuracies because line is the most clearly observable features, particularly in low resolution images, and thus has attracted most research efforts. For that, in our study, the Gabor filtering followed by a thresholding technique is used to extract the lines structure.

As mentioned, in our biometric system, the features are generated from the ROI sub-images by filtering it with 2D Gabor filter. In pattern recognition fields, this technique is widely used and it proves their efficiency vis a vis the simplicity of implantation and the system accuracies. Specifically, a 2D Gabor filter $G(x, y; \theta, \mu, \sigma)$ can be formulated as follows [10]:

$$G(x, y; \theta, \mu, \sigma) = \frac{1}{2\pi\sigma^2} e^{-\frac{1}{2}\left(\frac{x^2+y^2}{\sigma^2}\right)} e^{2\pi j \mu (x\cos\theta + y\sin\theta)} \tag{1}$$

where $j = \sqrt{-1}$, $\mu$ is the frequency of the sinusoidal signal, $\theta$ controls the orientation of the function, and $\sigma$ is the standard deviation of the Gaussian envelope. Thus, the response of a Gabor filter $(G(\theta, \mu, \sigma))$ to an image $(I)$ is obtained by a 2D convolution operation.

$$I_f(\theta, \mu, \sigma) = I * G(\theta, \mu, \sigma) \tag{2}$$

In our work, the $N \times N$ Gabor filter, $N = 16$, at an orientation, $\theta = \frac{\pi}{4}$, will convolute with the ROI sub-images. The results of a pair of a real and an imaginary filtered image are combined into a module response $\mathcal{A}$ as follows:

$$\mathcal{A} = \sqrt{\{Re(I_f(\theta, \mu, \sigma))\}^2 + \{Im(I_f(\theta, \mu, \sigma))\}^2} \tag{3}$$

The Gabor module response is encoded as "0" or "1" based on the sign. Therefore, the binary vector, $\mathcal{A}(i,j)$, is represented by the following inequalities:

$$\mathcal{A}_b(i,j) = \begin{cases} 1, & \text{if } \mathcal{A}(i,j) \geq 0 \\ 0, & \text{if } \mathcal{A}(i,j) < 0 \end{cases} \tag{4}$$

Finally, it is important to note that, in our series of experiments, the 2D Gabor filter parameters: $N$, $\mu$, $\sigma$ and $\theta$ are set as 16, 0.0916, 5.6179 and $\frac{\pi}{4}$, respectively.

## 3.2   Modeling Process

Generally, using a distance based classifier provide a significant classification error rate as well as an important times of features matching, especially, for the greater system database. Thus, in order to overcome these limits, classifiers based on modeling processes are used. Several model are used recently, but these based on deep learning techniques have been successfully provide a higher recognition rates. The following two sub-parts describe in detail each of the considered modeling processes which are the RBM model and the DBN model.

**Restricted Boltzmann Machine:** The RBM are presented by *Hinton* and *Sejnowski* as a special case of general Boltzmann Machine [11]. It is a stochastic generative neural network with stochastic binary units and undirected edges between units. Two types of units can be distinguished in RBM: visible units (observed variables) and hidden units (explanatory factors). These units are (distributed/arranged) in two layers [12]. So, the RBM has one layer of $I$ binary visible units, $v = [v_1, v_2, \cdots, v_I]$ where $v_i \in 0,1$, and a layer of $J$ binary hidden units, $h = [h_1, h_2, \cdots, h_J]$, where $h_j \in 0,1$, with bidirectional weighted connections. Thus, the RBM constitutes a network graph where each unit is connected to all the units in the other layer. While, the units of the same layer have no connections between them [13]. Figure 2 describes the structure of RBM.

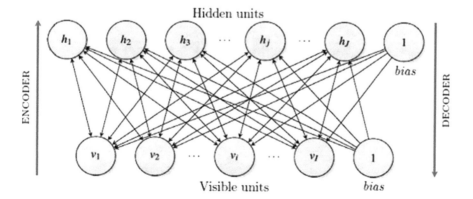

**Fig. 2.** Structure of the restricted Boltzmann machine (RBM).

The encoder process transforms the input into a feature vector representation. However, the decoder process reconstructs the original input from this feature vector. The joint probability distribution for both of visible and hidden units has an energy given by [14]:

$$P(v, h) = \frac{e^{-E(v,h)}}{Z} \tag{5}$$

where $Z$ is a normalization constant called partition function by analogy with physical systems, witch is given by summing over the energy of all possible pairs of visible and hidden vectors:

$$Z = \sum_v \sum_h e^{-E(v,h)} \tag{6}$$

The network assigns to a visible vector $v$ the probability that is obtained by summing over all possible hidden vectors $h$:

$$P(v) = \sum_h P(v, h) = \frac{1}{Z} \sum_h e^{-E(v,h)} \tag{7}$$

Then, the derivative of the log probability of a training vector with respect to a weight can be calculated as follows [15]:

$$\frac{\partial log P(v)}{\partial w_{ij}} = < v_i h_i >^0 - < v_i h_i >^\infty \tag{8}$$

where the angle brackets are used to denote expectations under the distribution specified by the subscript that follows. Then, a Contrastive Divergence algorithm is used in order to fasten the learning for a RBM [16]. The general idea is to update all the hidden units in parallel starting with visible units, then reconstruct visible units from the hidden units, and finally update the hidden units again (see Fig. 3). The learning rule is given by:

$$\Delta w_{ij} = \epsilon(< v_i h_i >^0 - < v_i h_i >^1) \tag{9}$$

where $\epsilon$ is a learning rate.

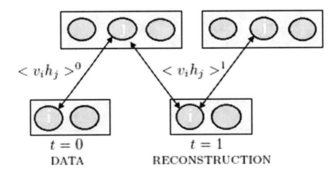

**Fig. 3.** Contrastive divergence algorithm for RBM.

**Deep Belief Networks:** In 2006, Hinton et al. [17] have introduced the DBNs witch are deep neural networks. They are a stack of RBM. Also, DBNs are composed of multiple layers of RBMs. Each layer is RBM and they are stacked each other to construct the DBN. The input enters the network in the first RBM. The output of the first RBM forms the input of the second, and so on until the actual output layer is reached [18] (see Fig. 4).

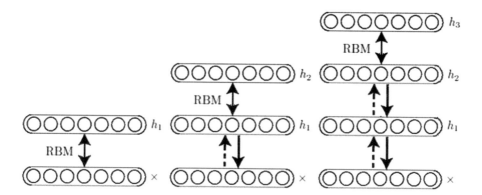

**Fig. 4.** Layer-wise training of a DBN.

## 4    Experimental Results

In our experiments, we use palmprint and palm-vein images dataset with size of 400 class (or persons), which is similar to the number of employees in small to medium sized companies. These dataset are extracted from multi-spectral palm-print database from the Hong Kong polytechnic university (PolyU) [19]. In each dataset, each person has twelve images for each modality. Thus, we randomly select three images for each modality in order to construct the system database (enrolment phase). The remaining nine images were used to test the system performance. However, by comparing nine test images with the corresponding class in the system database, we can obtain the client experiments. Similarly, by comparing these images with each class in the system database (except their class), we can obtain the impostor experiments.

### 4.1    Test Protocol

In our work, the set of experiments are divided into three sub-parts. In the first sub-part, we try to choose the optimal models parameters yield the best performance for RBM and DBN based classifier using the two biometrics modalities. The second sub-part focusing on the unimodal system performance using both PLP and PLV in the two modes of identification (open-set and closed-set identification). Note that, in these two sub-parts we use unimodal biometric systems. Finally, the last sub-part is devoted for evaluate the performance of the multimodal identification systems.

Thus, as mentioned above, our system is tested for the closed-set and open-set identification. In fact, in identification mode and when an unknown user is presented, the system must determine the identity of this user. It matches the acquired biometric characteristics with the models which include all the users enrolled in the database in order to make its decision according to the identification mode (closed set or open set). In closed-set identification mode, the class, that has the highest degree of similarity with the acquired biometric characteristics, is selected by the system as the identity of the user. In the other case (open set mode), the system indicate that the user presenting the acquired biometric characteristics is an enrolled user or not. Finally, it important to note that, the number of matching scores used to compute the different errors is equal 48150, where 3600 are genuine comparisons and the remaining, 44550, are impostor comparisons are generated.

## 4.2    Experimental Setup

Generally, in pattern recognition application such as biometric system, the classifier used has a greater impact on the recognition rate of the classification system. Because our system use RBM and DBN models which are depends on the number of hidden units, the first series of experiments is to selection the best number which give the best systems accuracies for the two models. Therefore, we test several number of hidden units (from 200 to 1200 units) using the PLP and PLV modalities in order to show the influence of this number on the biometric system. In Fig. 5, we compare the performance of system under different number of hidden units when the RBM model is used. Whereas, Fig. 5(a) shows the PLP based system performance, and Fig. 5(b) reports the PLV based system performance. From these Figures, it is clear that, firstly, RBM model is very effectively for recognize the PLP/PLV features which can give always an efficient identification rate. Secondly, a number of 600 units for both PLP and PLV provide an effectively performance with an Equal Error Rate (EER) equal to 0.046 % and a threshold $T_o$ equal to 0.8504 for PLP modality and EER of 0.058 % with $T_o$ of 0.8396 for PLV modality. It is important to note that, in this model (RBM), the feature vector is simply the entire image. So, our biometric systems can operate with a Genuine Acceptance Rate (GAR) equal to 99.954 % and 99.942 % for respectively PLP and PLV modalities.

Similarly, the same experiments are performed for the DBN model and the results are illustrated in Fig. 5(c) and (d) for PLP and PLV modalities. From these Figures, it is clear that, a number of 800 units for both PLP and PLV provide an effectively performance with a GAR equal to 99.969 % and a threshold $T_o$ equal to 0.8911 for PLP modality and GAR of 99.977 % with $T_o$ of 0.8692 for PLV modality. We also noted that, in this model (DBN), the feature vector is a binary template, for that our modalities are firstly filtered with Gabor filter and then binarized to obtain a binary feature. In conclusion, we choose these numbers for constructs open-set and closed-set identification biometric systems based on RBM and DBN models.

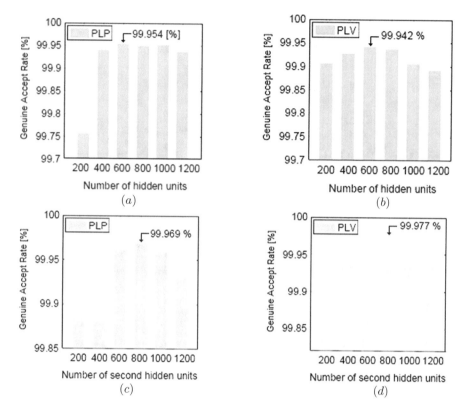

**Fig. 5.** Influence of the number of hidden units. *(a)* RBM based PLP identification (first hidden layer); *(b)* RBM based PLV identification (first hidden layer); *(c)* DBN based PLP identification (second hidden layer) and *(d)* DBN based PLV identification (second hidden layer).

### 4.3   Unimodal Biometric System Test Results

The objective of this sub-section is to compare the performance of the two deep learning methods (RBM and DBN) when are used as a classifier in a biometric systems based on PLP and PLV modalities. For that, our experiments results are also decomposed in two sets. The first set of results is concerning the open-set identification mode, whereas the second set of results is concerning the closed-set identification mode. For the open-set identification mode, Fig. 6*(a)* and *(b)* compare the systems performances under the two models for PLP and PLV modalities. From these Figure, it is clear that the DBN model works effectively than RBN. Thus, in this case a False Accept Rate (FAR) and False Rejected Rate (FRR) of 0.031 % and 0.023 % can be obtained for respectively PLP and PLV modalities. To present the closed-set identification experiments, we draw the Cumulative Match Characteristics (CMC) curves. This curve represents the identification rate (Rank-One Recognition (ROR)) against the Rank. Also,

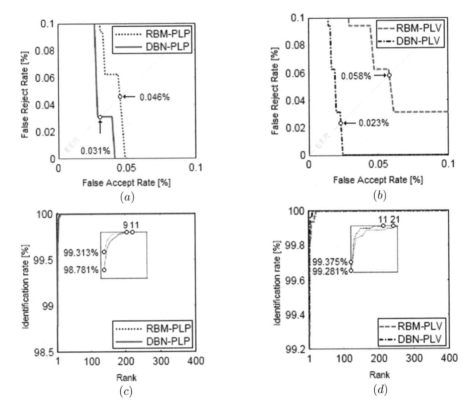

**Fig. 6.** Biometric system performances. *(a)* PLP based open set identification system; *(b)* PLV based open set identification system; *(c)* PLP based closed set identification system; *(d)* PLV based closed set identification system.

Fig. 6*(c)* and *(d)* show the superiority of using DBN than using RBM model. Thus, using the PLP modality, the proposed system achieves an identification rate equal to 99.313% (with a rank of perfect recognition equal to 11) when the RBM is used, and 98.871% (with a Rank of Perfect Recognition (RPR) equal to 9) when using DBN model. In the case of PLV modality, the use of DBN gives a ROR equal to 99.375% (with a RPR equal to 11). However, RBM model gives ROR of 99.281% (with a RPR equal to 21).

## 4.4   Multimodal Biometric System Test Results

Biometric system performance in the case of unimodal scenario produces some errors. For that, in this sub-part, we try to improve their performance by using the data fusion principal. However, in the proposed multimodal system, each modality, PLP & PLV, is operates independently and their results are combined using score level fusion scheme. In the two cases, open-set and closed-set identification modes, to find the better of the fusion rules (in our scheme, we examined five fusion rules which are: SUM, WHT (weighted sum), MUL (multiplication), MAX and MIN) an experimental result at the EER point as well as ROR point are shown in Tables 1 and 2. Table 1 relates to a comparison of the fusion rules by varying the classifier (RBM and DBN), in the open-set identification system. From this Table, it is clear that our open-set identification system achieves the best EER in all fusion rules with expected the MAX rule. However, it can operate at *ZeroEER* (EER = 0.000 %) in the case of MIN rule which is perfect recognition. Thuis, we have also performed a closed-set identification mode by applying all fusion rules on the matching scores obtained from the two modalities and evaluate the ROR and RPR as shown in Table 2. From this table, it can be seen that by combining PLP and PLV, the performance is in general improved. These results show that SUM, WHT, MUL and MIN fusion rules performs better than the MAX rule and improves the original performance (ROR equal to 99.969 % and RPR equal to 2). Finally, through a serious analysis of all obtained results, it can be concluded that in general the performance of the unimodal system is significantly improved by using the DBN model. In addition, these results are also demonstrated that multimodal fusion performs better results in the two modes of identification (open-set and closed-set).

**Table 1.** Multimodal *Open-set* identification test result (PLP and PLV fusion)

| Model | SUM | | WHT | | MUL | | MAX | | MIN | |
|---|---|---|---|---|---|---|---|---|---|---|
| | $T_o$ | EER | $T_o$ | EER | $T_o$ | EER | $T_o$ | EER | $T_o$ | EER |
| RBM | 0.9301 | $10^{-4}$ | 0.9221 | $10^{-4}$ | 0.8797 | $10^{-4}$ | 0.8943 | 0.0348 | $10^{-4}$ | $10^{-4}$ |
| RBM | 0.9361 | $10^{-4}$ | 0.9171 | $10^{-4}$ | 0.8940 | $10^{-4}$ | 0.9019 | 0.0312 | 0.9873 | 0.000 |

**Table 2.** Multimodal *closed-set* identification test result (PLP and PLV fusion)

| Model | SUM | | WHT | | MUL | | MAX | | MIN | |
|---|---|---|---|---|---|---|---|---|---|---|
| | ROR | RPR | ROR | RPR | ROR | RPR | ROR | RPR | ROR | RPR |
| RBM | 99.938 | 23 | 99.938 | 25 | 99.938 | 24 | 99.406 | 13 | 99.938 | 23 |
| DBN | 99.969 | 2 | 99.969 | 2 | 99.969 | 2 | 99.312 | 4 | 99.969 | 2 |

# 5  Conclusion and Further Works

In this paper, results obtained from palmprint and palm-vein modalities based biometric identification systems are provided. For that purpose, two methods of classification based on deep learning are used. Thus, the RBM and DBN model are used to classified the PLP and PLV and the experimental results, obtained on a database of 400 users, have shown a very high identification accuracy and proven the effectiveness of deep learning methods in the fields of biometric security. For further improvement, our future work will project to use other biometric modalities like finger-knuckle-print as well as the use of other deep learning methods. Thus, because of the various problems affecting the system performance of unimodal biometric based systems, and the proven effectiveness of multimodal biometric based systems, we will leave as future work to combine different unimodal systems using different fusion techniques.

# References

1. Wang, L.: Some issues of biometrics: technology intelligence, progress and challenges. Int. J. Inf. Technol. Manage. Inderscience Publisher **11**(1/2), 72–72 (2012)
2. Kanade, S., Petrovska-Delacrtaz, D., Dorizzi, B.: Multi-biometrics based cryptographic key regeneration scheme. In: IEEE 3rd International Conference on Biometrics: Theory, Applications, and Systems, BTAS 2009. IEEE (2009)
3. Chae, S.-H., Pan, S.B.: Improvement of fingerprint verification by using the similarity distribution. In: Information Technology Convergence, pp. 203–211. Springer, Dordrecht (2013)
4. Ross, A.A., Nandakumar, K., Jain, A.K.: Handbook of Multibiometrics, vol. 6. Springer Science & Business Media (2006)
5. Chergui, O., Bendjenna, H., Meraoumia, A., Chitroub, S.: Combining palmprint & finger-knuckle-print for user identification. In: IEEE International Conference on Information Technology for Organizations Development (IT4OD), Fez, Morocco, pp. 1–5 (2016)
6. Schmidhuber, J.: Deep learning in neural networks: an overview. Neural Netw. **61**, 85–117 (2015)
7. Koutnk, J., Schmidhuber, J., Gomez, F.: Evolving deep unsupervised convolutional networks for vision-based reinforcement learning. In: Proceedings of the 2014 Annual Conference on Genetic and Evolutionary Computation. ACM (2014)
8. Yang, J., Shi, Y., Yang, J.: Personal identification based on finger-vein features. Comput. Hum. Behav. **27**(5), 1565–1570 (2011)
9. Zhang, D., et al.: An online system of multispectral palmprint verification. IEEE Trans. Instrum. Meas. **59**(2), 480–490 (2010)
10. Wai, K.K., David, Z., Li, W.: Palmprint feature extraction using 2-D gabor filters. Pattern Recogn. **36**(10), 2339–2347 (2003)
11. Hinton, G.E., Sejnowski, T.J.: Learning and relearning in Boltzmann machines. Parallel Distrib. Process. **1** (1986)
12. Fresnel, Q., Peeters, G.: Apprentissage de descripteurs audio par deep learning, application pour la classification en genre musical. Dissertation. Univserity Paris, vol. 6 (2015)
13. Fischer, A., Igel, C.: Training restricted Boltzmann machines: an introduction. Pattern Recogn. **47**(1), 25–39 (2014)
14. Srivastava, N., et al.: Dropout: a simple way to prevent neural networks from overfitting. J. Mach. Learn. Res. **15**(1), 1929–1958 (2014)
15. Sarikaya, R., Hinton, G.E., Deoras, A.: Application of deep belief networks for natural language understanding. IEEE/ACM Trans. Audio Speech Lang. Process. (TASLP) **22**(4), 778–784 (2014)
16. Salakhutdinov, R., Mnih, A., Hinton, G.: Restricted Boltzmann machines for collaborative filtering. In: Proceedings of the 24th International Conference on Machine Learning. ACM (2007)

17. Hinton, G.E., Osindero, S., Teh, Y.-W.: A fast learning algorithm for deep belief nets. Neural Comput. **18**(7), 1527–1554 (2006)
18. Le Roux, N., Bengio, Y.: Representational power of restricted Boltzmann machines and deep belief networks. Neural Comput. **20**(6), 1631–1649 (2008)
19. Hong Kong Polytechnic University (PolyU) Multispectral Palmprint Database. http://www.comp.polyu.edu.hk/biometrics

# New Feature Extraction Approach
# Based on Adaptive Fuzzy Systems
# for Reliable Biometric Identification

Z. Tidjani[1], A. Meraoumia[2], S. Chitroub[3(✉)], and K. Ben sid[1]

[1] Laboratoire de Génie Électrique (LAGE),
Faculté des Nouvelles Technologies de l'Information et de la Communication,
Université Kasdi Merbah, Ouargla, Algeria
{Tidjani.zakaria,bensid.khaled}@univ-ouargla.dz
[2] LAboratory of Mathematics, Informatics and Systems (LAMIS),
University of Larbi Tebessi, 12002 Tebessa, Algeria
ameraoumia@gmail.com
[3] Laboratory of Intelligent and Communication Systems Engineering (LISIC),
Electronics and Computer Science Faculty, USTHB,
P.O. Box 32, El Alia, 16111 Bab Ezzouar, Algiers, Algeria
s_chitroub@hotmail.com

**Abstract.** Feature extraction for an optimal data representation is crucial for any biometric identification system. In this paper, we propose a new approach to extract the discriminant features within a biometric image in order to use Later in a biometric identification system. Thus, qualified as universal approximator, Takagi-Sugeno fuzzy system is adopted to model the biometric images through optimization of error target function, in which, a conjugate gradient method is used to establish the proposed algorithm. In order to evaluate our method, the PolyU multispectral palmprint database is used. The obtained results show that the biometric system errors are extremely reduced especially when the Blue spectral band is used. Thus, compared with the conventional features extraction methods, our method is more secure, fast and points at increased identification accuracy which will undoubtedly can be used in high secure applications.

**Keywords:** Biometric system · Multispectral palmprint
Takagi-Sugeno fuzzy system · Conjugate gradient · Data fusion

## 1 Introduction

In several applications, user identification is an essential foundation for protecting information. However, the need for enhancing this identification has expanded the research to focus on the biometric traits of the users. Thus, the biometrics offers a natural and reliable solution to the problem of identity

M. Alenezi and B. Qureshi (Eds.): 5th International Symposium
on Data Mining Applications, pp. 210–222, 2018.
https://doi.org/10.1007/978-3-319-78753-4_16

determination by recognizing personals based on their physical or behavioral characteristics [1]. Compared with traditional security means, the biometric-based security offers more properties and several advantages. So, among several biometric modalities, these extracted from the hand, *e.g.* palmprint, have been effectively used to make person identification for the last years. Palmprint has attracted an increasing amount of attention and it has proven to be a unique biometric identifier [2]. In the past few years, some researchers direct towards others features from the palm by using the multispectral imaging technology. Thus, the availability of some spectral bands in multispectral image has promoted research to build a multimodal system based on the fusion of these bands in order to overcome some of the limitations imposed by unimodal systems such as insufficient accuracy. In recent years, several researchers investigate application of fuzzy systems in biometric identification problems. Usually, fuzzy systems are introduced as classifier, as fusion operator or as decision module [3–6]. Others develop some specific fuzzy systems based on available previous experience information and observation. The fuzzy database rules are constructed on the basis of a linguistic formulation. Unfortunately, these works suffers from the lack of mathematical framework that ensures the efficiency of developed algorithms. A very useful property of fuzzy systems is their capacity to approximate unknown functions. Using a sufficient number of rules and an appropriate choice of fuzzy system parameters, it is possible to reproduce unknown function. To do that, it will be necessary to find a training algorithm to adjust correctly fuzzy system parameters. In this work, we propose a new feature extraction algorithm for biometric identification purpose. The characteristic vector is a collection of fuzzy database rule parameters obtained through optimization of error target function. For that, we will use conjugate gradient method as fuzzy model adaptation law. Thus, our contribution, is the introduction and development of adaptive fuzzy system as biometric image modeler with convergent feature extraction algorithm. A PolyU multispectral palmprint Database is used in the evaluation of the proposed unimodal and multimodal biometric system with KNN classifiers. The results of the classifier are then fused at the matching module. We have conducted a comparative study with other biometric systems to prove the robustness of our proposed system.

The remaining of the paper is organized as follows: the proposed system and the biometric associated function are briefly presented in Sect. 2. The image modeling with fuzzy system will be examined in Sect. 3. Our proposed feature extraction method will be developed in Sect. 4. For evaluation, experimental results, prior to fusion and after fusion, are given and commented with a comparative study in Sect. 5. Finally, the conclusion and some perspectives are presented in Sect. 6.

## 2 Biometric System Design

In our proposed biometric based identification system, the palmprint Region Of Interest (ROI) [2] is firstly extracted to be used as input training data to adaptive fuzzy system. Based on quadratic problem optimization, an off-line adaptive law is developed to look for some fuzzy optimal parameters that produce best

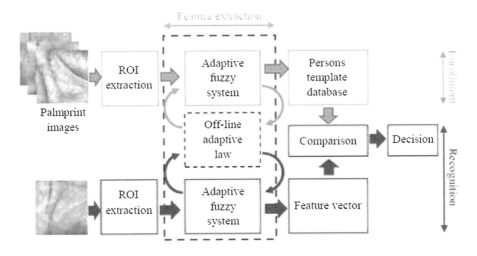

**Fig. 1.** Proposed biometric identification system.

approximation of image associated function. These parameters are proposed as feature vector of the palm print. As illustrated in Fig. 1, person template (or feature vector) are extracted in enrollment phase and stored in dedicated database. During identification phase and after the feature extraction process, the feature vector is compared to the previously stored vectors in database. The person will be accepted or rejected depending on a predefined security threshold. Thus, our main contributions in this study is the integration of adaptive fuzzy system as image modeling tool and the development of feature extraction problem as quadratic function optimization to be resolved with conjugate gradient method. As shown in the example of Fig. 2, the palmprint image is characterized by the grey level ($I$) of pixels in the image area. If the pixel position is described by the coordinates $(x,y)$ where $x$ and $y$ indicates respectively column and line pixel position, the grey level of image can be considered as nonlinear function of two variables such that:

$$I = I(x, y) \tag{1}$$

As illustrated in Fig. 3, the image can be considered as nonlinear function. In follows, we will construct an adaptive fuzzy system to approximate iteratively the

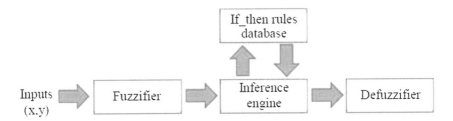

**Fig. 2.** Typical structure of fuzzy system.

image associated function. If done with sufficient accuracy, the fuzzy model as mathematical function will be considered an alternative to the biometric image to build person template.

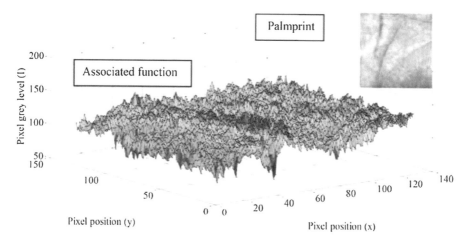

**Fig. 3.** Palmprint image associated function.

## 3   Fuzzy Systems Based Image Modeling

The heart of fuzzy systems is the if-then rules module. It regroups basic input-output relation to be used in output calculation for each input value. Integration of fuzzy systems in physical problem resolution can be done through two main visions: linguistic or abstract one according to nature of available information used to construct if-then rules. In the first one, the designer deals with relation between real world variables given as linguistic if-then rules to build an appropriate fuzzy system, when in the second one, as we will deal with, only numerical data are given and we have the task to define the appropriate rules to approximate target function. Basically, The fuzzy system is composed of four modules. In first, the input data is mapped to fuzzy space by fuzzifier then the result is inferred through database rules by the inference engine to produce fuzzy value as output. Finally, the deffuzzifier convert this value in numerical one.

In literature, Mamdani and Takagi-Sugeno models are two well known types of fuzzy systems [7]. They differ in the consequence of if-then rules. In the time Mamdani model use linguistic consequence, the Takagi-Sugeno models prefers numerical function. In this work, we will interest only with Takagi-Sugeno Fuzzy Model (TSFM) with two inputs and one output. Let us consider, TSFM with the input vector $X = [x, y]$ and the output $I_f$ where the pixel position coordinates $x$ and $y$ belongs to the two subspaces $\Omega_X$ and $\Omega_Y$, respectively. We define $n_x$

and $n_y$ fuzzy sets on $\Omega_X$ and $\Omega_Y$ respectively to form $m$ fuzzy rules where $m = n_x.n_y$. The $l^{th}$ rule can be expressed as follows

$$R^l : \text{if } x \text{ is } A^{i_l} \text{ and } y \text{ is } B^{j_l} \text{ then } I_f = f^l(x,y)$$

Where $A^{i_l}$ and $B^{j_l}$ are fuzzy sets associated to variables $x$ and $y$ respectively and characterized by the fuzzy membership functions $\mu_{A^{i_l}}(x)$ and $\mu_{B^{j_l}}(y)$ with $i_l \in \{1, \cdots, n_x\}$ and $j_l = \in \{1, \cdots, n_y\}$. The functions $f^l(x,y)$ are user defined function, usually chosen as polynomials to shape the fuzzy output. If the fuzzy system processing is based on singleton fuzzifier, product inference engine and center -average defuzzifier, the output will be expressed as the aggregation of all $m$ rules such that [7]

$$I_f(x,y) = \frac{\sum_{l=1}^{m} f^l(x,y).\mu_{A^{i_l}}(x).\mu_{B^{j_l}}(y)}{\sum_{l=1}^{m} \mu_{A^{i_l}}(x).\mu_{B^{j_l}}(y)} \tag{2}$$

In the particular case of zero order polynomial functions $f^l(x,y) = \theta^l$, the TSFS output can be rewritten in affine form with respect to parameters $\theta^l$ as follows:

$$I_f(x,y) = \theta^T \xi(x,y) \tag{3}$$

where $\theta = [\theta^1, \theta^2, \cdots, \theta^m]^T$ is a vector of adjustable parameters and the fuzzy basis function vector $\xi = [\xi^1, \xi^2, \cdots, \xi^m]^T$ is defined as:

$$\xi^l(x,y) = \frac{\mu_{A^{i_l}}(x).\mu_{B^{j_l}}(y)}{\sum_{k=1}^{m} \mu_{A^{i_k}}(x).\mu_{B^{j_k}}(y)} \tag{4}$$

Since the vector $\xi(x,y)$ can be computed using arbitrarily well defined membership functions as gaussian or triangular, the fuzzy system will be completely defined if we find an appropriate vector $\theta$ such that $I_f$ in (3) is sufficiently close to the corresponding image associated function given by (1).

Mathematical developments prove the existence of at least one vector that permits to fuzzy system output to reproduce an unknown function with sufficient accuracy, if all its variables are available [7]. This main property, called universal approximation, make the TSFM very suited in modeling and identification. Let us give the following statement [8]:

**Lemma 1.** Let $F(X)$ be a continuous function that is defined on compact set $\Omega_X$. For any given positive constant $\epsilon_{max}$, there always exits a fuzzy logic system $I_f(X)$ in the form of (3) such that:

$$sup_{X \in \Omega_X} |F(X) - I_f(X)| = |F(X) - \theta^T \xi(X)| < \epsilon_{max} \tag{5}$$

the optimal parameters vector $\theta^*$ is defined as [8]

$$\theta^* = argmin_{\theta \in \Omega_\theta}[Sup_{X \in \Omega_X}|F(X) - \theta^T \xi(X)|] \tag{6}$$

Lemma 1 confirms the existence of an optimal parameters vector, and allow us to approximate the image associated function as follows:

$$I(x,y) = \theta^T.\xi(x,y) + \epsilon(x,y) \tag{7}$$

$\epsilon(x,y)$ is considered as error modeling that strong depends on both TSFM parameters and the function to approximate. The maximal error $\epsilon_{max}$ should be reduced to an acceptable level by increasing particularly the number of input fuzzy sets. In this last equation, is a suggestion to construct an new image model using TSFM. Unfortunately, it give us no solution on how to completely define an optimal parameters vector.

## 4    Proposed Feature Extraction Algorithm

The TSFM is considered as an alternative representation of the corresponding biometric image. We will adopt unified fuzzy structure with the same fuzzy sets. If we examine, Eq. (3), TSFM of all images will share the same fuzzy basis functions vector $\xi(x,y)$ and the difference will reside only in the parameters $\theta$ that we suggest as the feature vector.

### 4.1    Quadratic Optimization Problem

In this subsection, we will deal with the requirement that parameters vector has to fulfill in the case of image modeling as preliminary step to construct the feature extraction algorithm. To obtain an optimal parameters vector, it is necessary to take into account approximation errors on all image pixels. The average of errors square, as one of favorite performance indexes is considered as measures of fuzzy approximation quality of one palmprint image. It is Let us define the error target function as

$$E(\theta) = \frac{1}{N_p} \sum_{(x,y)} (I(x,y) - I_f(x,y))^2 \tag{8}$$

where $N_p$ denote the number of pixels in image and the sum term concerns with all image pixels $((x,y) \in \Omega_x \times \Omega_y)$. Using Eq. (3) the error target function can be rewritten in the following compact form:

$$E(\theta) = \theta^T A\,\theta + b^T\theta + c \tag{9}$$

where the symmetric matrix $A \in R^m \times R^m$ only depends on the fuzzy structure, the vector $b \in R^m$ merges informations from both fuzzy system and palmprint image grey level and finally the scalar $c \in R_+$. They are given by

$$\begin{cases} A & = \dfrac{1}{Np} \sum_{(x,y)} \xi(x,y).\xi(x,y)^T \\[2mm] b^T & = -\dfrac{1}{Np} \sum_{(x,y)} 2I_f(x,y).\xi(x,y)^T \\[2mm] c & = \dfrac{1}{Np} \sum_{(x,y)} I_f(x,y)^2 \end{cases} \tag{10}$$

The error expression given by (9) is formulated in a second order quadratic form with respect to the unknown vector $\theta$. To resolve it, we will develop a variant of the well known gradient optimization algorithm.

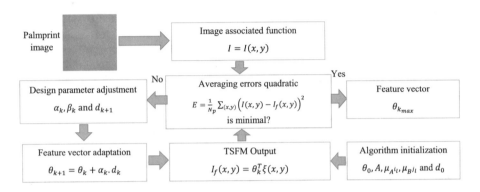

**Fig. 4.** The proposed feature extraction algorithm.

## 4.2    Conjugate Gradient Method

Basic gradient method proposes an iterative algorithm to resolve linear and nonlinear optimization problem. In this method, an arbitrary fixed step is used to compute a new solution vector towards the opposite side of the gradient of the error target function. A basic algorithm is given by [9]:

$$\theta_{k+1} = \theta_k - \eta.\nabla_{\theta_k} E(\theta_k) \tag{11}$$

where the integer $k$, the positive scalar $\eta$ and the operator $\nabla_{\theta_k}$ denotes iteration index, step size parameter and gradient function vector respectively. The conjugate method is a variant that proposes a modified structure as given by the following equation [9]

$$\theta_{k+1} = \theta_k + \alpha_k.d_k \tag{12}$$

The new search direction line $d_k$ is given as

$$d_{k+1} = \nabla_{\theta_k} E(\theta_k) + \beta_k.d_k \tag{13}$$

At each iteration, the variable length step $\alpha_k$ is adapted to guarantee the error energy optimization according to the actual parameter vector. It must minimizes in-posteriori error given by

$$E(\theta_{k+1}) = \theta_{k+1}^T.A.\theta_{k+1} + b^T.\theta_{k+1} + c \tag{14}$$

If we replace (12) in the last equation, the optimal value of $\alpha_k$ is easily obtained by equating with zero its derivative. We find

$$\alpha_k = -\frac{d_k^T(A.\theta_k + b)}{d_k^T.A.d_k} \tag{15}$$

In Eq. (13), the scalar design sequence $\beta_k$ is introduced to construct successive search lines in conjugate directions. The two vector $d_{k+1}$ and $d_k$ will be conjugate with respect to the matrix $A$ if $d_{k+1}.A.d_k = 0$. Since, from (9), the gradient is given by $\nabla_{\theta_k} E(\theta_k) = A.\theta_k + b$ and taking into account the relation (13), the previous condition is satisfied with the following adaptation law

$$\beta_k = \frac{(A.\theta_k + b)^T.A.d_k}{d_k^T.A.d_k} \tag{16}$$

The proposed algorithm, as shown in Fig. 4, can be summarized as follows:

1. Choose fuzzy system: $n_x, n_y, \mu_{A^{i_l}}(x)$ and $\mu_{B^{j_l}}(y)$
2. Compute matrix $A$ and vector $b$ from Eq. (10)
3. Initialize vectors $\theta_0$, $d_0 = -(A.\theta_0 + b)$ and index $k = 0$;
4. Compute:
   (a) The new step size $\alpha_k$ use Eq. (15)
   (b) New solution $\theta_k$ given by Eq. (12)
   (c) Adaptive parameter $\beta_k$ see Eq. (16)
   (d) Direction vector $d_{k+1}$ use Eq. (13)
5. Compute the error given by (14) if not small enough, increase $k = k + 1$ return to step (4).
6. Record the final solution $\theta_k$ as the feature vector.

## 5   Experimental Results and Discussion

We experimented and evaluated our approach on hong kong Polytechnic University (PolyU) multispectral Palmprint Database [10]. The database has a total of 4800 images obtained from 400 persons. The images were taken in two different sessions separated by a time interval of about two months. During each period, each individual had to take at least six pictures of his palmprint. Moreover, in the second period, the light source and the (CCD) camera lens are adjusted so that the images of the first and second session gave the impression of having been taken by two different captured devices. Images have also been taken under different light conditions to test the robustness of the biometric system. However, the extracted ROI sub-image size is equal to $128 \times 128$ with a resolution of 75 $dpi$. Finally, we note that the acquisition system collects four images from four bands (Red, Green, Blue and NIR). To develop a palmprint identification system, it is necessary to take two databases: a database to perform learning and other one to test and determine their performance. There are no rules that fixes the manner to share one available database into two parts. In our series of tests, we have divided the multispectral palmprint database as follows:

**Learning Images:** the first, fifth and ninth image of each person (for each spectral band) to serve learning phase.

**Table 1.** Identification rate under the design parameters.

| $k_{max}$ | 5 | 10 | 15 | 20 | 25 | 30 | 35 | 40 |
|---|---|---|---|---|---|---|---|---|
| Identification rate (%) | 97.77 | 97.88 | 98.27 | 98.66 | 99.02 | 99.36 | 96.25 | 96.52 |
| $n_x=n_y$ | 5 | 10 | 15 | 20 | 25 | 30 | 35 | 40 |
| Identification rate (%) | 68.41 | 99.36 | 99.55 | 99.61 | 99.63 | 99.66 | 99.63 | 99.63 |

**Test Images:** The remaining 9 images of each individual have helped us achieving different tests.

After that, we will first discuss the influence of fuzzy parameters design in the proposed method and we chose the best of these. Then we will examine the impact of the proposed algorithm in case of uni-modal biometric system followed by the case of multi-modalities.

The GAR (Genuine Accept Rate) is important index for measuring the performance of biometric system, it means the rate of customers accepted by the biometric system. Fuzzy membership functions are in triangular form and equally distributed on the $x$ and $y$ intervals. The proposed algorithm contains two impor-

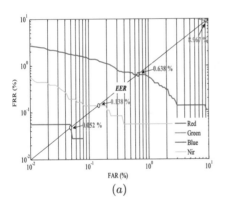
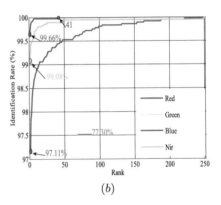

(a)    (b)

**Fig. 5.** Unimodal identification test results. *(a)* Unimodal open-set identification test results, *(b)* Unimodal closed-set identification test results.

**Table 2.** Unimodal identification results

| Spectral bands | Open-set identification | | Closed-set identification | |
|---|---|---|---|---|
| | $T_o$ | EER | ROR | RPR |
| Red | 0.142 | 0.638 | 97.11 | 248 |
| Green | 0.125 | 8.967 | 77.30 | 394 |
| Blue | 0.129 | 0.052 | 99.66 | 41 |
| NIR | 0.126 | 0.138 | 99.08 | 250 |

tant design parameters: the number of iterations $k_{max}$, as algorithm stop condition, and the number of membership functions ($n_x$ and $n_y$). For this, we have studied in the first step influence of the two parameters in the permeability rate of the biometric system. Biometric identification can be applied to open set or closed set of persons. In the first case, the system allows to add anyone even if it is outside the database. However, in closed set identification, the biometric system recognizes only the people in the database. The Table 1 illustrates the GAR against the number of both iteration and membership functions. In this Table, we note that the biometric system gives values greater than 99.00% for $k_{max}$ equal 25, 30 but the best is obtained with 30 it gives 99.36% from the GAR. All values below 25 give good values with less processing time. $k_{max}$ values greater than 30 give lower results than their predecessors, take much longer time and require the use of high-impact devices. Now we hold the value of $k_{max}$ at 30 and see the influence of the number of the membership functions. From the previous table we note that all values except 5 give good results (greater than 99.36%) but the perfect is obtained with $n_x = n_y = 30$. When this number is greater than 30 it takes more computing time, which require special intensive calculator equipment, without recording enhancement. At the end, and after the comparison of the previous results, we conclude that best parameters of proposed method are obtained with $n_x = n_y = 30$ and $k_{max} = 30$.

## 5.1   Unimodal System Result

To evaluate the identification system performance, we will use the information from each spectral band of the palmprint modality. For this, in open-set identification, we found the performance under different spectral bands (*Red, Green, blue, NIR*). Figure 5 and Table 2 compares performance of the unimodal system using fuzzy feature extraction for different spectral bands. If we consider the Equal Error rate (*EER*), the rate at which both acceptance and rejection errors are equal, the experimental results indicate that the *Blue* perform better than the NIR, *Red* and *Green* in (*EER*). It gives an $EER = 0.052\%$ with threshold $T_0 = 0.129$. But in the general case, the rest of bands give relatively good results. The *EER* is less than 0.700% except for the *Green* spectral band as shown in Fig. 5(a). In the case of closed-set identification system, Fig. 5(b) shows the Cumulative Match Characteristic (*CMC*) curve which represent the identification rate against all ranks. In this figure, all bands show good results, except the *Green* band. The rate of identification at first test, known as the Rank-One Recognition (ROR), is superior to 97.000% for *Blue, Red, NIR* bands and the perfect result is obtained with the *Blue* band since it gives the greatest value 99.660% and from the Rank of Perfect Recognition *RPR* where the identification achieves 100%, the *Blue* band give a best results when it gives 41 and the *Green* give a very bad value estimated at 77.30% from *ROR* and 394 from *RPR*.

## 5.2   Multimodal System Result

Unimodal system showed good performance, but we aspire to obtain more perfect results, so there are several ways to improve the performance of the unimodal system. But the famous are achieved through multimodal biometric system. To enhance identification performance, we have increased the amount of discriminate information of each person by combining several modalities, including what comes from the different sensors, which is called multi-sensor, the multi-algorithm when it uses different feature extraction method. Also, we find multi-samples where various samples are used for the same biometric modality and the multi-biometric based on the integration of information from different biometric modality. In this part, several unimodal systems are combined at matching score level. For that, different fusion rules were tested (*SUM, WHT, MIN, MAX* and *MUL*) to find the rule that optimizes the system accuracy. Scores of subsystems should be normalized before applying the fusion. This normalization is performed using the *min-max* scheme [11]. In our multimodal identification system, the scores of some sub-systems are combined, at the matching score level, to generate unique score which will be used to make the final decision. Thus, Table 3 shows the performance of the multimodal identification system for, respectively, open-set and closed-set identification modes. These results indicate that: in open-set identification all the fusion method give a better result compared to the unimodal results but the fusion of two bands give a best than the fusion of three bands ($EER = 0.048\%$). Also, same results are given in the closed-set identification ($ROR = 99.66\%$). In this case, The Blue-Nir fusion results outperform those of Red-Blue and Red-Green bands. Furthermore, the *Sum* and *Mul* rules are the best rules compared to all other rules since they gives results without

**Table 3.** Multimodal identification systems test results (EER %)

| Open set identification | | | | | | | | |
|---|---|---|---|---|---|---|---|---|
| Fusion rules | SUM | | MUL | | MAX | | MIN | |
| | $T_o$ | EER | $T_o$ | EER | $T_o$ | EER | $T_o$ | EER |
| Red-Green | 0.145 | 1.133 | 0.010 | 0.359 | 0.176 | 2.277 | 0.045 | 0.441 |
| Red-Blue | 0.129 | 0.048 | 0.016 | 0.055 | 0.171 | 0.111 | 0.062 | 0.063 |
| Red-Green-Blue | 0.137 | 0.115 | 0.069 | 0.098 | 0.213 | 0.944 | 0.012 | 0.111 |
| Blue-Nir | 0.189 | 0.05 | 0.028 | 0.038 | 0.190 | 0.055 | 0.115 | 0.055 |
| Closed set identification | | | | | | | | |
| Fusion rules | SUM | | MUL | | MAX | | MIN | |
| | ROR | RPR | ROR | RPR | ROR | RPR | ROR | RPR |
| Red-Green | 94.66 | 239 | 87.33 | 273 | 87.33 | 338 | 93.00 | 381 |
| Red-Blue | 99.66 | 128 | 98.52 | 112 | 98.52 | 78 | 99.5 | 183 |
| Red-Green-Blue | 99.05 | 123 | 87.47 | 27 | 87.47 | 27 | 96.52 | 327 |
| Blue-Nir | 99.86 | 106 | 99.25 | 92 | 99.25 | 70 | 99.86 | 171 |

**Table 4.** Performances comparison

| Methods | Papers | Accuracy (EER %) |
|---|---|---|
| Variant of Local Binary Patterns (LBP) | [12] | 1.133 (best band) |
| 2D Gabor phase coding scheme | [13] | 0.748 (best band) |
| Histograms of oriented gradients (HOG) | [14] | 0.358 (best band) |
| Fusion code | [15] | 0.615 (fusion band) |
| Complete direction representation | [16] | 0.430 (best band) |
| Our | | 0.052 (best band) |
| | | 0.028 (best two bands) |

additional computing complexity which confirm the improvement effect of the fusion in identification accuracy. A comparison study with other feature extraction methods is performed. For that purpose, we consider as references those based on Local Binary Patterns [12], on 2D Gabor phase coding scheme [13], on Histograms of Oriented Gradients (HOG) [14], on feature level fusion [15] and on Complete Directional Representation [16]. Table 4 summarizes the obtained results in which we observe that the proposed method result outperforms those of the previously cited works. Indeed, our biometric system gives best results for both unimodal and multimodal identification systems using the KNN classifier. With accuracy level near to 0.028% the EER is extremely reduced to prove the efficiency of the proposed feature extraction algorithm.

# 6   Conclusion and Further Works

In this paper, we have proposed a new feature extraction method based on Takagi-Sugeno fuzzy model. As fuzzy system parameters, the feature vector is obtained iteratively by optimizing a suggested error target function through conjugate gradient method. We had incorporated this method in a biometric identification system that we have developed. We have compared the obtained results with the results of the other methods developed in the literature. According to the evaluation criteria of biometric identification systems, our method has greatly exceeded other methods. For further improvement, we project in our future work to improve the effectiveness of our method other research fields such as: image segmentation and biomedical image processing.

# References

1. Kekre, H.B., Sarode, T., Vig, R., Arya, P., Bisani, S., Irani, A.: Identification of multi-spectral palmprints using energy compaction by hybrid wavelet. In: IEEE International Conference on Biometric, New Delhi, India, pp. 433–438 (2012)
2. Meraoumia, A., Kadri, F., Chitrob, S., Bouridane, A.: Improving biometric identification performance using PCANet deep learning and multispectral palmprint. In: Biometric Security and Privacy-Opportunities and Challenges in The Big Data Era, pp. 51–69. Springer (2017)

3. Ghayoumi, M., Ghazinour, K.: An adaptive fuzzy multimodal biometric system for identification and verification. In: 2015 IEEE/ACIS 14th International Conference on Computer and Information Science (ICIS), Las Vegas, NV, pp. 137–141 (2015)
4. Monwar, M.M., Gavrilova, M., Wang, Y.: A novel fuzzy multimodal information fusion technology for human biometric traits identification. In: 2011 10th IEEE International Conference on Cognitive Informatics and Cognitive Computing (ICCI*CC), Banff, AB, pp. 112–119 (2011)
5. Vasuhi, S., Vaidehi, V., Babu, N.T.N.: An efficient multi-modal biometric person authentication system using Fuzzy Logic. In: ICoAC 2010, Chennai, pp. 74–81 (2010)
6. Wang, L.X.: Fuzzy systems are universal approximators. In: 1992 Proceedings of the IEEE International Conference on Fuzzy Systems, San Diego, CA, pp. 1163–1170 (1992)
7. Ying, H., Ding, Y., Li, S., Shao, S.: Comparison of necessary conditions for typical Takagi-Sugeno and Mamdani fuzzy systems as universal approximators. IEEE Trans. Syst. Man Cybern. Part A Syst. Hum. 29(5), 508–514 (1999)
8. Li-Bing, W., Yang, G.-H., Wang, H., Wang, F.: Adaptive fuzzy asymptotic tracking control of uncertain nonaffine nonlinear systems with non-symmetric dead-zone nonlinearities. Inf. Sci. 348, 1–14 (2016)
9. Conjugate gradient methods. In: Numerical Optimization. Springer Series in Operations Research and Financial Engineering. Springer, New York (2006)
10. PolyU Palmprint database. http://www4.comp.polyu.edu.hk/biometrics/
11. Yan, X., Kang, W., Deng, F., Wu, Q.: Palm vein recognition based on multisampling and feature-level fusion. Neurocomputing 151(2), 798–807 (2015)
12. Ojala, T., Pietikäinen, M., Mäenpää, T.: Multiresolution gray-scale and rotation invariant texture classification with local binary patterns. IEEE Trans. Pattern Anal. Mach. Intell. 24(7), 971–987 (2002)
13. Zhang, D., Kong, W., You, J., Wong, M.: Online palmprint identification. IEEE Trans. Pattern Anal. Mach. Intell. 25(9), 1041–1050 (2003)
14. Dalal, N., Triggs, B.: Histograms of oriented gradients for human detection. In: Proceedings of the IEEE Computer society Conference on Computer Vision and Pattern Recognition (CVPR), pp. 886–893 (2005)
15. Kong, A., Zhang, D., Kamel, M.: Palm print identification using feature-level fusion. Pattern Recognit. 39(3), 478–487 (2006)
16. Jia, W., Zhang, B., Lu, J., Zhu, Y., Zhao, Y., Zuo, W., Ling, H.: Palmprint recognition based on complete direction representation. IEEE Trans. Image Process. 29, 4483–4498 (2016)

# An Efficient PHSW-DC Algorithm for Solving Motif Finding Problem in TP53 Cancer Gene

Asmaa G. Seliem[1], Wael Abouelwafa[2]([✉]) $\bullet$, and Hesham F. Hamed[1]

[1] Electrical Engineering Department, Faculty of Engineering,
Minia University, Minya, Egypt
[2] Bio-Medical Engineering Department, Faculty of Engineering,
Minia University, Minya, Egypt
wael.wafa@mu.edu.eg

**Abstract.** Bioinformatics scientists are interested in a computational tools that identifies frequency occurring substrings (motifs), which hypothetically have a significant function in whole genome for Deoxyribonucleic Acid (DNA) sequence analysis. This paper proposed a novel algorithm based on Smith Waterman algorithm using the technique of divide and Conquer (HPSW-DC). Both software and hardware accelerators are introduced to accelerate the motif finding algorithms. This paper implements this algorithm on Field Programmable Gate Array (FPGA), which needs hardware specialists to design such systems. A parallel tri-sequence technique decrease the resource utilization, improves the accuracy, increase the computation throughput and accelerate the performance moreover, it enables an alignment for big data available for a complete gene. The proposed algorithm had been synthesize using Xilinx ZynQ7000 which gave us 43 ns execution time, 14% utilization at 869.944 MHz frequency and 72 GCUPS for tetra-nucleotide.

**Keywords:** DNA · Motif · PHSW-DC · ZYNQ7000

## 1 Introduction

Genetic information in living organism store in DNA sequences. A DNA motif is a nucleic acid sequence pattern that has a biological significance such as being DNA binding sites for a regulatory protein. Normally, the motif pattern is short (2 to 20 base-pairs) what's more, is known to repeat in various genes or a few times inside a gene [1]. Motif finding in DNA sequences is a challenging task in molecular biology. Many algorithms was created for discovering motifs and each of motif has points of interest for research.

Motifs occur repeatedly in the sequence with mutations in some of their nucleotide positions. DNA motifs often represent Transcription Factor Binding Sites (TFBS) where Transcription Factors proteins bind at these sites to regulate the expression of genes hence discovery of such motifs helps to understand the mechanisms of gene expression [2].

M. Alenezi and B. Qureshi (Eds.): *5th International Symposium on Data Mining Applications*, pp. 223–233, 2018.
https://doi.org/10.1007/978-3-319-78753-4_17

Computational motif finding has emerged with the computational biology and become a promising studied area of research because of its importance [3].

The motif finding issue is a nondeterministic polynomial-time (NP) problem, so there is no polynomial-time solution present for motif finding. A challenge task is to pick out regulatory elements, especially the binding sites in (DNA) for transcription factors [4].

For instance, when a fly becomes infected, the dormant immunity genes in the fly genome activated and start producing proteins that will destroy the pathogen. The regulatory motifs activate this gene, so it is importance for scientists to knew what these motifs are and how they functioning. The best approach that is currently known is to research many patterns of DNA and search for patterns due to the fact the patterns is probably the motifs in question.

Regulatory motifs control the expression or regulation of a group of genes involved in a similar cellular function. Identification of these motifs gives insight into the regulatory mechanism of genes. Motif finding algorithms purpose to find out those commonplace patterns, which may additionally present in the DNA. Identity of DNA motifs is complex due to mutations, which lead them to conserved patterns [5].

In this paper, an FPGA-based architecture designed to accelerate the motif finding using a developed algorithm based on Smith waterman (SW). The significance of the study is to improve the performance, optimize the power consumption used, increase speed and decrease utilization.

The rest of the paper is organized as follows; the Sect. 2 describes literature survey about motif finding and used algorithms in short. Section 3 describes the proposed system. Section 4 gives the result of paper. Section 5 gives conclusion of the paper.

## 1.1 Motif Finding Problem and Proposed Solution

A planted motif search (PMS) is a method for identifying conserved motifs within a set of nucleic acid sequences, it is a widely addressed problem in many literatures and simplified combinatorial problem of biological The PMS problem was first introduced by Keich and Pevzner [6] and described as follows: if there are n number of sequences of length t, every and every sequence is planted with an instance of the accord motif M of length m each having at the most d mutations, confirm the accord motif. Motif accord is that the motif that doesn't have any mutations.

The scientific literature describes numerous algorithms for solving the PMS problem. These algorithms can be classified into two types. Heuristic algorithms that return the approximate solution as Random Projection [7], Pattern Branching [8]. However, most algorithms yield an actual continued time to break the planted motif search problem (PMS).

Some of PMS used is microsatellite, a microsatellite is an adjacent repeated motifs that range in length up to five nucleotides, and are typically repeated up to 50 times. For example, the sequence TACGTACGTACG is a Tetra-nucleotide microsatellite, and GTACTTGTACTT is a Hexa-nucleotide microsatellite (with A being Adenine, G Guanine, C Cytosine, and T Thymine). Microsatellites are distributed throughout the genome.

There are many problems facing motif finding algorithms as:

- huge data in databank,
- which lead to large time for algorithm for execution
- And accuracy due to the size of motif it's self.

So we need fast and accurate algorithm to solve these problems.

Software based acceleration techniques have been used and introduced in Grid Computing [9].

Nowadays multicore CPUs are getting used to accelerate many applications that require massive amount of data. The Multiple Expectation Maximization for Motif Elicitation (MEME) is a popular and efficient approximate algorithm used to accelerate motif finding. Then, hardware acceleration mechanisms such as using Field Programmable Gate Arrays (FPGA) and Graphics Processing Units (GPU) have been introduced. Chen et al. [10]. Farouk et al. [11] have implemented a new version of the brute force motif finding algorithm known as Skip-Brute Force search on FPGA.

## 1.2 Smith Waterman Algorithm

This proposed system based on smith waterman algorithm (SW) based on dynamic programming (DP) which is the most accurate local algorithm is implemented in a new technique (divide and conquer) to use SW semi-globally and locally in same time.

Smith Waterman algorithm steps:

(i)   Initialization and calculation matrix.
(ii)  Find high score in matrix.
(iii) Trace back alignment process.

Create matrix N*N dimension which m length of reference and query sequences, during the initialization and calculation matrix step, the first row and column are filled with zeros then, the matrix cells will be calculated and filled using Eqs. (1) and (2), where $H_{i,j}$ the elements of matrix, $S_{i,j}$ is the similarity score of the comparison between query sequence (Q) and reference sequence (R) and D is penalty of mismatch.

Figure 1 show smith waterman steps where (a) initialization and filling matrix process, (b) calculation cell by align two opposite letter one from reference and one from query the output of by comparing the left, right, diagonal cells and zero. The dynamic programming (DP) computes a matrix whose elements cannot be computed before computation of its north, northwestern and west elements [12]. After matrix calculation complete the final back tracing process from the highest score to lowest one, there may be several paths back to the original point. The generation of results indicates the outgoing sequences after alignment. The alignment with highest sum-up score is the best alignment DP computes a matrix.

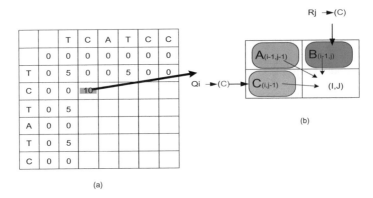

**Fig. 1.** (a) Smith Waterman matrix, (b) cell calculation depend on upper, left and diagonal cells.

$$H_{(i,0)} = H_{(0,j)} = 0 \tag{1}$$

$$H_{(i,j)} = max \begin{cases} H_{(i-1,j-1)} + S_{(i,j)} \\ H_{(i,j-1)} - D \\ H_{(i-1,j)} - D \\ 0 \end{cases} \tag{2}$$

### 1.3  Divide and Conquer (D&C) Technique

Divide and conquer is a useful technique for solving complex problems: by breaking the problem into sub-problems, and combining their solutions to be the final solution for original problem. This technique helps in the discovery of efficient algorithms D&C technique leads to an improvement in the asymptotic cost of the solution. The cost of the divide-and-conquer algorithm will be O(n log pn) where n problem's size and p number of sub problems.

### 1.4  Smith-Waterman Algorithm Pseudo Code

The pseudo code of Smith Waterman algorithm split into three component.

Inputs: reference sequence, query sequence, value of gap penalty, values of match and mismatch, dimension of matrix, Process: the main algorithm function which create matrix, initializing first raw and column then calculate every cell by using Eq. (2) and Output: give results of alignment. And combine them.

**Smith waterman Algorithm:**

-------------------------------------------------------------------------

INPUT:

Strang A:  Query sequence of length n to be aligned
Strang B: Reference sequence of length m to aligned
Gap extension =8, Gap insertion penalty =10
Match =5, MisMatch = -4
SWArray: n*m internal alignment Matrix

Process:

Step 1: initialization & score Matrix

SWAraay = zeros

Loop_I : for I =1:lenA
Loop_J:for j=1:lenB

IF  strA(i)= strB(j)

    SWArray(I,j)= SWArray(i-1,j-1) + match
End

IF strA (i)/= strB(j)

    SWAraay(I,j)=SWArray(i-1,j-1)+MisMatch
End

End loop
End loop

Step 2 : Highscore // find highest score in matrix

For i=1:lenA
  For j= 1:lenB
    IF (SWArray(I,j) >Hiscore)
    Hiscore = SWArray(I,j)
    HIpos(1)=i
    Hipos(2)=j
  End
 End
  End
OUTPUT:

OptA: aligned Reference sequence
OptB: aligned Query sequence

## 2   Method

### 2.1   Parallel Hardware Smith-Waterman Algorithm with Divide and Conquer Technique (PHSW-DC)

Parallel Hardware smith-waterman algorithm with divide and conquer technique (PHSW-DC) is proposed as a parallel accelerator for DNA motif finding. This Proposed architecture was designed in Very High Speed Integrated Circuit Hardware Description language VHSIC HDL (VHDL), it was then synthesized for Xilinx Zynq-7000 [13].

This technique for DNA motif finding (PMS) introduced as flow:

(a)   Assign each nucleotide in 2 bit size for faster alignment.
(b)   Break-down the DNA reference sequence(R) to P subsequence where p number of processes, the length of this subsequence equal the length of motif this break done by using method of divide and conquer; to reduce the complexity of the main structure and to parallel the process of alignment.
(c)   Each subsequence aligned with motif in parallel process.
(d)   Finally conquer all outputs of processes to get result of PMS.

The PHSW-DC algorithm is a local-global algorithm for sequence alignments, for performance acceleration and more optimal solution of the alignment so for simplicity, if we have a reference sequence from gene banks with a short read query sequence Q, we divide the reference sequences to sub-sequences equal to query sequence in length $R1:R1-1, R1-2, R1-3, \ldots R1-n$ as shown in Figs. 2 and 3, then we starts to compute Smith Waterman matrix in parallel for every sub-sequence with the short read query in same time, then compute the trace back step in matrices in parallel finally we combine the final results for all nine alignment in three final combination to get local-global final alignment to the short read query.

**Fig. 2.**   PHSW-DE algorithm block diagram

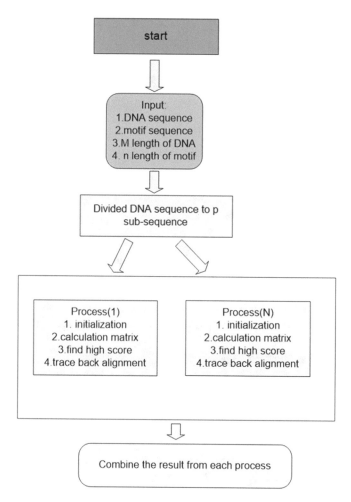

**Fig. 3.** Design flow chart of PHSE-DE algorithm.

## 2.2 Hardware System Design

A complete VHDL code was implemented for parallel smith-waterman algorithm using divide and conquer method to accelerate motif finding in ZYNQ 7000.

The design was synthesized from VHDL for Xilinx Zynq-7000 AP SoC XC7Z020-CLG484 FPGA Ver.D. Using Xilinx software tools (ISE 14.7) and simulated in ISim.

As shown in Fig. 4 each nucleotide in motif and reference sequences are compared, in a comparator, the result matched or mismatched value is send to multiplexer, which it is output added to the value of diagonal cell, then get the maximum as a final output with upper cell plus gap, left cell plus gap and zero. Then output of each processes conquer and get final result. Counter use to get address of subsequences reference that store in memory which will compare with motif.

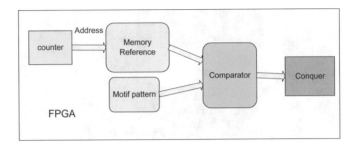

**Fig. 4.** Hardware design inside FPGA.

## 3   Complexity

In normal algorithm the complexity of the exhaustive search is O (Lnt) where L is the length of the motif, n is the length of the TP53 cancer gene, and t is the number of DNA samples so the approach takes exponentially longer to remedy as more DNA sequences are checked HPSW-DE algorithm we implemented has a complexity of O(Lnt/p) where p is the number of parallel processes. So the exaction time is decreased by a factor of parallel process P as it is clearly displayed in the simulation process shown in Fig. 5.

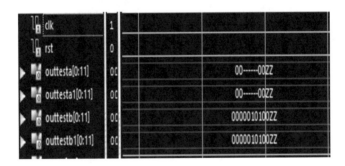

**Fig. 5.**   Simulation result of tetra-nucleotide alignment of PHSW-DC.

## 4   Results and Discussions

In this proposed system planted motif finding of Tetra, Penta and Hexa-nucleotide has been applied by PHSW-DC and implemented on ZYNQ 7000 FPGA. Figures 5 and 6, show simulation result of Tetra, Penta and Hexa-nucleotide alignment.

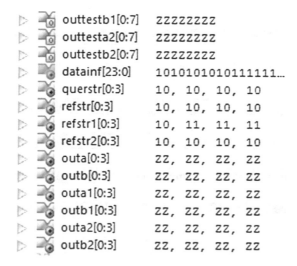

| | | |
|---|---|---|
| ▷ | outtestb1[0:7] | zzzzzzzz |
| ▷ | outtesta2[0:7] | zzzzzzzz |
| ▷ | outtestb2[0:7] | zzzzzzzz |
| ▷ | datainf[23:0] | 1010101010111111... |
| ▷ | querstr[0:3] | 10, 10, 10, 10 |
| ▷ | refstr[0:3] | 10, 10, 10, 10 |
| ▷ | refstr1[0:3] | 10, 11, 11, 11 |
| ▷ | refstr2[0:3] | 10, 10, 10, 10 |
| ▷ | outa[0:3] | zz, zz, zz, zz |
| ▷ | outb[0:3] | zz, zz, zz, zz |
| ▷ | outa1[0:3] | zz, zz, zz, zz |
| ▷ | outb1[0:3] | zz, zz, zz, zz |
| ▷ | outa2[0:3] | zz, zz, zz, zz |
| ▷ | outb2[0:3] | zz, zz, zz, zz |

**Fig. 6.** Simulation inputs and outputs of tetra-nucleotide alignment of PHSW-DC.

The Performance measurement for new PHSW-DC algorithm is calculated with the Eq. (3):

$$GCUPS = \frac{m \times n}{T \times 10^{-9}} \tag{3}$$

Where m and n are the sizes of both reference sequences and motif respectively and T is the execution time in seconds, The FPGA synthesize report with Speed Grade: −3 declare that minimum period: 1.149 ns (Maximum Frequency: 869.944 MHz), Minimum input arrival time before clock: 0.576 ns, Maximum combinational path delay: No path found. Table 1 presents the execution times, GCUPS, frequency and LUT utilization of the FPGA.

**Table 1.** ZYNQ 7000 utilization, execution time, GCUPs and efficiency.

| | Tetra-nucleotide | Penta-nucleotide | Hexa-nucleotide |
|---|---|---|---|
| GCUPs | 72 | 64.6 | 41.7 |
| Time (ns) | 43 | 60 | 111.8 |
| Utilization | 14% | 41% | 57% |
| Efficiency | 1 | 0.34 | 0.25 |

## 5  Conclusion

DNA alignment with a motif finding problem has a long history and there exist a number of efficient algorithms. This paper provided an enhancement of the Smith-Waterman algorithm allowing alignment of motif sequence Tetra to Hexa (3 to 6 letters alphabet) named HPSW-DC using Xilinx ZYNQ-7000 FPGA.

The SW algorithm is the most accurate in performing sequence alignment, however it requires an exceptionally long time for solving motif finding problem, making it the most appropriate alternative for hardware acceleration The new version for SW algorithm has been adopted to FPGA architecture which resulted in good parallelization and efficient pipelining. Presented benchmarks show that our implementation can run on a modern FPGA chip over high speed.

The proposed algorithm had been synthesize using Xilinx ZynQ7000 which gave us 43 ns execution time, 14% utilization at 869.944 MHz frequency and 72 GCUPS for tetra-nucleotide.

For TP53 cancer gene, which is a faster execution compared to the standard algorithms for motif finding problem compiled on a PC workstation. The SW algorithm without divide and conquer technique and parallel processes cannot be implemented in the same FPGA, because it will exceed the FPGA LUT's in the future this algorithm can be implemented in cluster FPGA's to solve the motif finding problem in hole genome.

**Acknowledgment.** The authors would like to thank Prof. Kamel H. Rahouma, Department of Electrical and Communication Engineering, minia University, Egypt for his support of this study.

# References

1. Rombauts, S., Déhais, P., Van Montagu, M., Rouzé, P.: PlantCARE, a plant cis-acting regulatory element database. Nucleic Acids Res. **27**, 295–296 (1999)
2. Das, M.K., Dai, H.-K.: A survey of DNA motif finding algorithms. BMC Bioinform. **8**, 1 (2007)
3. Perera, P., Ragel, R.: Accelerating motif finding in DNA sequences with multicore CPUs. In: 2013 IEEE 8th International Conference on Industrial and Information Systems, pp. 242–247 (2013)
4. Chauhan, R., Agarwal, P.: A review: applying genetic algorithms for motif discovery. Int. J. Comput. Technol. Appl. **3**, 1510–1515 (2012)
5. Seeja, K.: AISMOTIF-an artificial immune system for DNA motif discovery. arXiv preprint arXiv:1107.1128 (2011)
6. Keich, U., Pevzner, P.A.: Finding motifs in the twilight zone. In: Proceedings of the Sixth Annual International Conference on Computational Biology, pp. 195–204 (2002)
7. Buhler, J., Tompa, M.: Finding motifs using random projections. J. Comput. Biol. **9**, 225–242 (2002)
8. Price, A., Ramabhadran, S., Pevzner, P.A.: Finding subtle motifs by branching from sample strings. Bioinformatics **19**, ii149–ii155 (2003)
9. Faheem, H.: Accelerating motif finding problem using grid computing with enhanced brute force. In: The 12th International Conference on Advanced Communication Technology (ICACT), pp. 197–202 (2010)
10. Chen, C., Schmidt, B., Weiguo, L., Müller-Wittig, W.: GPU-MEME: using graphics hardware to accelerate motif finding in DNA sequences. In: IAPR International Conference on Pattern Recognition in Bioinformatics, pp. 448–459 (2008)
11. Farouk, Y., Faheem, H., ElDeeb, T.: Massively Parallelized DNA Motif Search on FPGA. INTECH Open Access Publisher (2011)

12. Steinfadt, S.I.: SWAMP+: enhanced Smith-Waterman search for parallel models. In: 2012 41st International Conference on Parallel Processing Workshops (ICPPW), pp. 62–70 (2012)
13. Seliem, A.G., El-Wafa, W.A., Hamed, H.F.A.: Hardware acceleration of Smith-Waterman algorithm for short read DNA alignment using FPGA. In: Presented at the COMPSAC 2016: The 40th IEEE Computer Society International Conference on Computers, Software & Applications, Atlanta, Georgia, USA (2016)

# The Effect of Vitamin B12 Deficiency
# on Blood Count Using Data Mining

Nada Almugren$^{(\boxtimes)}$ ⓘ, Nafla Alrumayyan ⓘ, Rabiah Alnashwan ⓘ,
Abeer Alfutamani ⓘ, Isra Al-Turaiki, and Omar Almugren ⓘ

King Saud University, Riyadh, Saudi Arabia
nada.almugren@hotmail.com, naflaru@gmail.com,
rabiahalnashwan@gmail.com

**Abstract.** Healthcare systems create vast amount of data collected from medical examination. Data mining techniques are widely used in healthcare systems to detect diseases in early stages. In this paper, we applied four data mining techniques to find the relation between vitamin B12 levels and blood cell count. Four data mining techniques were applied to real patients' dataset: Neural Networks (MLP), Naïve Bayes, J48, and JRip. The resulting models were evaluated using the real datasets obtained from King Khalid University Hospital (KKUH), Riyadh, Saudi Arabia. Experimental results showed that both MLP and JRip techniques were capable of classifying the dataset correctly regardless of the size of the dataset.

**Keywords:** Healthcare · Data mining · Data mining techniques · Neural networks
Naïve Bayesian · J48 · JRip · Vitamin B12

## 1 Introduction

In recent years, health care systems produce an extensive and complex data from medical examinations. Finding specific rules, patterns, and trends in the extensive database is considered significant. Moreover, in the medical field the data contains unobserved patterns, which give the ability to find out diseases, and in best case is to predict diseases before it occurs.

Data mining was found to be an efficient method for analytical study of background factors in blood stream infection [15]. As a matter of fact, many researchers find out that applying different types of data mining techniques and methods have the ability to detect diseases in an early stage. B12 is an essential vitamin and is one of the eight B vitamins family. The deficiency of B12 leads to a condition called *Macrocytic anemia* (reduction in all blood cell).

The aim of this study is to use data mining techniques to identify the relation between vitamin B12 level measured in the lab and blood cell count including white blood cells (WBC) count, platelets count, hemoglobin level and red cell indices (Mean Corpuscular Volume (MVC), Mean Corpuscular Hemoglobin (MCH), Red Cell Distribution Width (RDW), Retics). The data was obtained from King Khalid University Hospital (KKUH),

© Springer International Publishing AG, part of Springer Nature 2018
M. Alenezi and B. Qureshi (Eds.): *5th International Symposium*
*on Data Mining Applications*, pp. 234–245, 2018.
https://doi.org/10.1007/978-3-319-78753-4_18

Riaydh, Saudi Arabia. The dataset contains all patients' records that had vitamin B12 level measured for the period from November 2016 to October 2017. We applied four data mining classifier techniques to the dataset, including: Neural Networks (MLP implemented in Weka) [1], Naive Bayes [2], J48 decision tree [3], and JRip [4] rule induction algorithm. Performance evaluation was then done based on *accuracy, error rate, precision, recall, and f-score* measures. Acceptable performance was observed for all techniques. However. MLP and JRip techniques achieved 100% accuracy. This means that MLP and JRIP techniques are capable of classifying all dataset correctly regardless of the size of the training dataset. In addition, MLP and JRip show the perfect results of F-score in all their experiments results.

The rest of the paper is organized as follows: the second section discusses the literature review; the third section addresses the methodology, the fourth section about experiments and results discussion. Finally, the last section denotes the conclusion and future works of the paper.

## 2   Literature Review

There is growing research interest called medical data mining which is concerned systems that predict and extract knowledge from medical data. data mining techniques are being used to investigate several types of hematology diseases. It is indubitably that laboratory information is extremely important in many different fields. Therefore, it is crucial to increase the efficiency and the quality of diagnosis. This would be achieved through studying and exploring the relationship between laboratory results and diseases.

Saichanma et al. [5] used J48 decision tree to predict abnormality in peripheral blood smear. The authors analyzed a dataset that includes 1362 teenagers' records and by using 13 data set of hematological parameters gathered from automated Blood Cell analyzer (SysMex XT1800i, Sysmex corporation, Kobe, Japan). Those records are classified into two groups: normal RBC morphology and abnormal RBC morphology. 25.99% of the dataset was abnormal RBC morphology (no imbalance mentioned). The author used 6 measures to evaluate the performance of J48: TP, FP, precision, recall, F-measure, and accuracy which gave results of 0.940, 0.050, 0.945, 0.940, 0.941, and 0.943, respectively.

In a study by a Abdullah and Al-Asmari [6], The authors predicted the most common five anemia types through conducting data mining classification algorithms. The dataset they used was Complete Blood Count (CBC) test results, the authors filtered attributes by eliminating the irrelevant attributes from the dataset. Four classification algorithms were used Naïve Bayes [2], neural network (Multilayer Perception) [1], J48 decision tree and Support Vector Machin. Authors found that and J48 decision tree performs best accuracy with 93.75% in the percentage split 60%. Also, they found that J48 decision tree gives the best performance with F-Measure, Sensitivity, The true positive rate and precisions. Thus, J48 proved to be the most effective and efficient classification algorithm in this context. The dataset of this study was only of 41 patients which is small dataset.

Another study that aims to predict and classify anemia types using data mining techniques [7]. The dataset was used contains Complete Blood Count (CBC) tests of 514 patients. They used three classification methods: (C4.5) decision tree algorithm and support vector machine which are implemented as J48 and Sequential Minimal Optimization (SMO). Their aim is to find the best method for prediction and best possible classification of anemia. Evolution result showed that C4.5 algorithm is best classification technique with the highest accuracy 99.42%.

In addition, a new clustering algorithm (weight-based k means) developed by Vijayarani and Sudha [8]. the authors utilized existing data mining techniques to predict diseases from the hemogram blood test. In the study, Leukemia, Inflammatory disease, bacterial or viral infection, HIV infection and Pernicious anemia diseases were predicted by using 3 different techniques: K means clustering algorithm, fuzzy c means clustering [9] and the weight based k means clustering algorithms. Those techniques were evaluated for their performances using the clustering accuracy, error rate and execution time, where weight-based k means clustering algorithm got the highest score in term of accuracy and time.

Ferraz et al. presented a prototype for the determination of blood pre-transfusion tests [10]. The authors are interested in reducing the test time of blood pre-transfusion. The prototype used two approaches for blood classification: Decision Trees and support vector machines. The prototype software uses image processing and classifiers to identify the result of blood type test. Unlike the previous approaches of using image processing, this prototype provides an approach where different types of Regions of Interest (ROIs) are considered (square ROIs and circular ROIs) to detect agglutination, and thus determine the blood type.

Classification has also been applied in order to diagnose genetic disease. In paper [11] they investigated Thalassemia genetic diagnosis using three data mining classifiers: Decision Tree, Naïve Bayes, and Neural Networ. Thalassemia has a direct influence on a hemoglobin in red blood cells. The authors mainly depend on Complete Blood Count (CBC) data to distinguish between Iron Deficiency patients, Thalassemia patients or other blood diseases. They conducted different experiments and resulted that the Naïve Bayes is most significant classifier with the percentage of accuracy exceeding 90%, and most indication features Thalassemia existence is Mean Cellular Volume(VCM) with a value less than 77.65.

Authors in [12] also depends on CBC attributes (WBC1, RBC2, Platelets, Hemoglobin…etc.) to discover related diseases to the patient blood. However, CBC is susceptible to noise, so they rely on the data pre-possessing includes data refining, transformation, and data recoding. Their aim is grouping the peoples who share the same kind of diseases by using clustering algorithms. The efficiency of the clustering was improved by using data normalization based on PLT3 attribute to undergo the data within small scale of range. Then inconsistency also has been resolved through data recoding technique which is converted Blood Group (BG) data format to numeric format to ensues CBC data completely refined. k-mean algorithms were used for clustering with Expectation maximization (EM) to find an accurate number of clusters. The authors conducted two experiments and they found that pre-processed data produce an efficient and optimized result.

Platelets k-mean algorithm has been widely used among with clustering algorithms. Paper [13] use the K-Means Algorithms and Generation of Association Rules on Blood Cell Counter Data from a clinical laboratory. Twelve thousand records resulted from an automated machine which is called Automated Blood Cell Counter generate Complete Blood Count (CBC) features as spreadsheet, such as RBC count, WBC count, Mean Cell Volume (MCV) platelets count of a given blood sample. These reports were transformed and refined using KDD pre-processing phases. Then they applied Clustering and Association Rules on useful data. Association Rules results shown that the Apriori algorithm results are more useful than general Association Rule Mining algorithm. Then they applied K-Means algorithms based on RBC attribute with the various value of k and they get different numbers of clusters each time.

The paper [14] shows how they applied data mining techniques to discover the relations between blood test characteristics and blood tumor in order to predict the disease in an early stage. The aim of this study is to use data mining techniques to classify CBC sample of a blood disease patient as normal hematology disease or blood tumor. The dataset has been used in this paper contains 5350 samples of patients divided into two groups, group one labeled as "Tumor" has contains 1764 samples while the other group labeled as "Hematology" contains 3586 samples. In addition, this paper conducted three experiments using three different types of classifiers: classification based association rules, rule induction and deep learning. First, as association rule is used to investigate which CBC test has a relation with blood tumor sample. Second, rule induction is used to discover patterns that associated with blood tumor and normal hematology classes and the results come with accuracy 71.66%. Third, the results of deep learning come with accuracy 79.45%, however, the result has no explanation. As can be seen, the deep learning classifier has the best ability to detect tumor from blood samples disease.

The paper [15] shows the data mining analysis of relationship between blood stream infection and clinical background in patients undergoing lactobacillus therapy. The aim of this study is to analyze the impacts of lactobacillus therapy and the background risk factors of blood stream infection in patients by using data mining techniques. Moreover, they used the clinical data collected from patients and the population for this study consists of 1291 patients. In addition, the analytical methods used are: decision tree, chi-square test and logistic regression.

First, "if-then rules" shows the relationship between the bacteria detection and the various factors. Also, decision tree showed that bacteria have the strong relationship with lactobacillus therapy, diarrhea, catheter, and tracheotomy. Second, chi-square test was applied to evaluate the association between lactobacillus therapy and blood stream infection (bacteria detection on blood cultures) and these results showed that lactobacillus therapy might have the effect to the prevention of blood stream infection. Third, logistic regression was applied to analyze the relationship between bacteria detection and lactobacillus therapy, anti-biotics, etc. Also, the result of the logistic regression showed that bacteria detection has strong relationship with lactobacillus therapy, diarrhea, etc. As a result, researchers consider to take the more specific measures for the prevention of hospital infections based on these analytical results.

# 3   Methodology

## 3.1   Data Collection

The main interest of this paper is to identify the relation between vitamin B12 level and *Macrocytic anemia* (reduction in all blood cell) through blood cell count, platelets count, and hemoglobin level and red cell indices. To do that the study will analyze a data that contain all previously mentioned measures.

The data collection process was based on two stages; finding a hospital that could provide vitamin B12 data associated with patients' information, which is in our case King Khalid University Hospital. Second, describing the desired data to the hospital DBA in order to extract the right data from the hospital database. The extracted data contain vitamin B12 tests and blood count of patients in a period of 12 months from November 2016 to October 2017. The data set consists of 5637 records, its attributes represent general patient information, Vitamin B12 result, and Complete Blood Count (CBC) test features in Table 1.

**Table 1.** Vitamin B12 result, and Complete Blood Count(CBC) test features

| Attribute name | Attribute description | Data type |
|---|---|---|
| Sex | Gender of the patient | Binominal |
| Age-Years | Age (Visit) | Integer |
| B12 | Vitamin B12 result | Real |
| Hct | Hematocrit | Real |
| Hgb | Hemoglobin | Real |
| MCH | Mean corpuscular hemoglobin | Real |
| MCHC | Mean corpuscular hemoglobin concentration | Real |
| MCV | Mean corpuscular volume | Real |
| MPV | Mean platelet volume | Real |
| Plt | Platelet count | Real |
| RBC | Red blood cell | Real |
| RDW | Red cell distribution width | Real |
| UWBC | Urine white blood cells | Real |
| WBC | White blood cell | Real |

## 3.2   Data Pre-processing

Real data is noisy, inconsistent, and incomplete which needs a data pre-processing to make it clean and ready to use. Data pre-processing is a data mining technique and a significant phase in data mining process. There are many tasks in data pre-processing such as data cleaning, data integration, data transformation, data reduction and data discretization. In our dataset, we did some data pre-processing. For instance, first, some of the attributes were not columns; therefore, we used java code to solve this problem which converts the row attributes to column. Second, we mostly have missing values in age attributes. To deal with this kind of problem, we use marginal imputation methods which compute the mean from the non-missing values and use the mean to replace the missing values in the dataset [14]. Moreover, some of the useless attributes were eliminated like: Order Date & Time, MCHC (Mean corpuscular hemoglobin concentration), MPV (Mean platelet volume) and UWBC (Mean corpuscular volume). Finally, we have 24 files and each file contains the data for 15 days. After we filled the missing values, we combined the 24 files into one file, which eventually contains 4213 records.

In data transformation, we change the data type of some attributes to the appropriate format for analysis. The CBC features and B12 result converted to the nominal data type with three labels: High, Normal and Low. Diagnosis attribute also added which represents a class label of the data with three labels: Anima, No Anima, and Macrocytic Anima. All data labeling made by a physician from King Khalid University Hospital depending on medical reports of the laboratory. The resulting data was class imbalanced, so we have overcome this issue by using over-sampling and under-sampling. SMOTH approach was applied for over-sampling and SpreadSubSample for the under-sampling [17]. Table 2 below defines the ranges of CBC features and B12 of different ages and sex and the diagnosis attribute values.

**Table 2.** The ranges of CBC features and B12 of different ages and sex and the diagnosis attribute values.

| Test | Sex | Age | Range | |
|------|-----|-----|-------|-------|
| WBC | | 2 years | 5 | 15.5 |
| | | 6 years | 4.5 | 13.5 |
| | | 15 years | 4 | 11 |
| RBC | | 3 years | 4.0 | 5.5 |
| | M | 15 years | 4.7 | 6.1 |
| | F | 15 years | 4.2 | 5.5 |
| Hgb | | 1 years | 105 | 135 |
| | | 6 years | 120 | 140 |
| | | 3 years | 115 | 145 |
| | M | 15 years | 130 | 180 |
| | F | 15 years | 120 | 160 |
| HCT | | 3 years | 35 | 45 |
| | M | 15 years | 42 | 52 |
| | F | 15 years | 37 | 47 |
| MCV | | 2 years | 77 | 85 |
| | | 6 years | 80 | 91 |
| | | 15 years | 50 | 94 |
| MCH | | 15 years | 27 | 32 |
| RDW | | | 11.5 | 14.5 |
| PLT | | | 140 | 450 |
| B12 | | | 145 | 637 |

### 3.3  Classification Methods

For the purpose of this study, WEKA is used to evaluate the dataset of 4213 instances by using 4 data mining classifier technique: Neural Networks (MLP in Weka), Naive Bayes, J48 and JRip.

A decision tree is a set of attributes shaped as tree-structured design to predict the output and shows the outcome from series of decisions. J48 is a Java implementation of C4.5 decision tree algorithm in Weka data mining tool. J48 creates a decision tree based on labeled data using information entropy. Then, it can be tested on unseen data to define how well the model performed. The decision trees generated by J48 can be used for classification. We choose J48 because it is accuracy and its ability to extract both decision tree and rule-sets, and construct a tree [16].

JRip (RIPPER) Implements a propositional rule learner proposed by W. Cohen as an optimized version of IREP. It is one of the most popular and basic algorithms where the classes are examined in an initial set of rules. Then proceeds these rules by training all the examples of a decision in the training data as a class. After that, finding a set of rules that cover all the members of that class. Then proceeds to the next class until all classes covered [4].

Multilayer perceptron is a type of Artificial Neural Networks (ANN) with back-propagation algorithm. MLP is an inter connected network of artificial neurons (nodes). The behavior of neurons is simulating the behavior of a human brain. Therefore, it provides a mean to solve complicated problems. A multilayer perceptron network contains an input layer that consists of source neurons, one or many hidden layers of computation neurons, and an output layer of neurons. The input signal propagates between the network till it reaches the result [18].

The Naive Bayes classification is a simple and effective probabilistic classifier technique that is based on Bayesian theorem with strong independence (naive) assumptions. Where it learns the probability of an object with certain features belonging to a particular group in a class. It represents statistical and supervised learning algorithms. The Naive Bayesian is called "naive" because it determines the probabilistic occurrences of certain events is independent of the occurrences of other events [19].

## 4  Experiments and Results Discussion

In this paper, four classification techniques naïve Bayes, J48 (C4.5), neural networks (MLP) and JRip were used and compared on basis of accuracy, Error Rate, Precision, Recall and f-measure. Tenfold cross-validation and percentage split (60%, 80%, 90%) of the dataset were used in the experiment. The results in Table 3 show the evaluation of the dataset using WEKA through different experiments and measurements. The experiments use percentage splits that are conducted by choosing random percent of data for training and the rest of data for testing.

**Table 3.** Evaluation the dataset using WEKA through different experiments and measurements

|  | Training set | Accuracy% | Mean absolute error% | Precision | Recall | F-measure |
|---|---|---|---|---|---|---|
| Multilayer perceptron | 60% | 100% | 0.007 | 1.000 | 1.000 | 1.000 |
|  | 80% | 100% | 0.006 | 1.000 | 1.000 | 1.000 |
|  | 90% | 100% | 0.0005 | 1.000 | 1.000 | 1.000 |
|  | Cross validation | 100% | 0.006 | 1.000 | 1.000 | 1.000 |
| Naïve Bayes | 60% | 98.87% | 0.0082 | 0.989 | 0.989 | 0.989 |
|  | 80% | 99.88% | 0.0067 | 0.999 | 0.999 | 0.999 |
|  | 90% | 100% | 0.0037 | 1.000 | 1.000 | 1.000 |
|  | Cross-validation | 99.29% | 0.0072 | 0.993 | 0.993 | 0.993 |
| J48 | 60% | 99.8% | 0.0014 | 0.98 | 0.98 | 0.98 |
|  | 80% | 99.6% | 0.0025 | 0.97 | 0.97 | 0.97 |
|  | 90% | 99.7% | 0.0015 | 0.98 | 0.98 | 0.98 |
|  | Cross validation | 99.9% | 0.0006 | 0.99 | 0.99 | 0.99 |
| JRip | 60% | 100% | 0 | 1.00 | 1.00 | 1.00 |
|  | 80% | 100% | 0 | 1.00 | 1.00 | 1.00 |
|  | 90% | 100% | 0 | 1.00 | 1.00 | 1.00 |
|  | Cross validation | 100% | 0 | 1.00 | 1.00 | 1.00 |

Accuracy (Correctly Classified Instances) is an effective measure of providing the overall performance of an algorithm, through showing the correctly classified true values of class label. The accuracy is computed as follows:

$$accuracy = \frac{tp + tn}{tp + fp + fn + tn} \tag{1}$$

Table 3 Shows the accuracy of all experiments in this study. Although all experiment got decent accuracy rates, MLP and JRIP techniques achieve 100% accuracy in all data split percentages and tenfold cross-validation. This means that MLP and JRIP techniques are capable of classifying all dataset correctly regardless of the size of the training dataset. While Naïve Bayes achieves 100% accuracy in only data split of 90%.

Precision is called the positive predictive value, it is the ratio of positive predictions to the total predicted positive class. Precision is the measure of the classifier exactness which computed as follows:

$$precision = \frac{tp}{tp + fp}; \tag{2}$$

High precision means low false positive rate. We have got 1.00 precision on two algorithms JRip and neural networks (Multilayer Perceptron) that's mean all classes predicted by these algorithms are correct. The other two algorithms have more than 0.97 precision which is very good. Recall is called sensitivity or the true positive rate, it is he ratio of positive predictions to the number of all observations in actual class which computed as follows:

$$recall = \frac{tp}{tp + fn} \tag{3}$$

JRip and neural networks (Multilayer Perceptron) got recall of 10.00 that's mean all predicted classes were correct. The other two have got recall 0.97 which is good.

F-score is also used in this study to test out the experiments' results. F-score is considered as a harmonic mean of precision and recall (see Eq. 4). A higher value of f-score is preferable because it implies more perfect values of precision and recall. Based on these considerations, we can conclude from Table 3 that MLP and JRIP provide perfect results of F-score in all their experiments.

$$F - measure = \frac{(\beta^2 + 1) * precision * recall}{\beta^2 * precision + recall} \tag{4}$$

Figure 1 show the result of the J48 classifier, where we can notice all type of anemia is mainly dependent on Hgb. Table 4 shows the result of JRip rule classifier. Accordingly, Macrocytic Anemia depends on the result of four attributes MCH, RDW, MCV, and Hgb while the other kinds of anemia are depending only on two attributes Hgb and Hct.

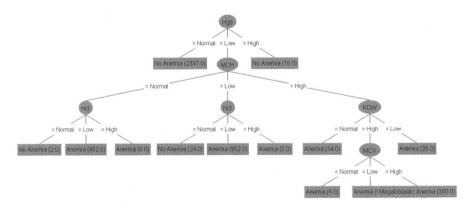

**Fig. 1.** The result of the J48 classifier

**Table 4.** Result of JRip rule classifier

| JRip rule Result |
| --- |
| (MCH = High) and (RDW = High) and (MCV = High) and (Hgb = Low) => Diagnosis = Macrocytic Anemia |
| (Hgb = Low) and (Hct = Low) => Diagnosis = Anemia |
| => Diagnosis = No Anemia |

Figure 2 represent the neural network of our dataset. The network illustrates 3 different kinds of neurons (nodes) the output neurons (in yellow) represent whether the patient is normal, has anemia or has a Macrocytic anemia, the hidden neurons (in red), and the input neurons (in green) represents categories of each attribute.

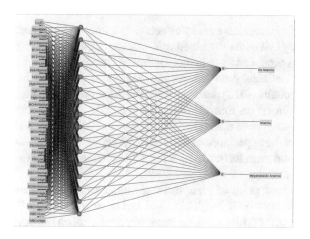

**Fig. 2.** Represent the neural network of our dataset

| Attribute | Class No Anemia (0.56) | Anemia (0.35) | Megaloblastic Anemia (0.09) |
|---|---|---|---|
| **B12** | | | |
| Normal | 2103.0 | 1264.0 | 365.0 |
| Low | 159.0 | 124.0 | 5.0 |
| High | 91.0 | 102.0 | 9.0 |
| [total] | 2353.0 | 1490.0 | 379.0 |
| **Hgb** | | | |
| Normal | 2308.0 | 1.0 | 1.0 |
| Low | 27.0 | 1488.0 | 377.0 |
| High | 18.0 | 1.0 | 1.0 |
| [total] | 2353.0 | 1490.0 | 379.0 |
| **MCH** | | | |
| Normal | 1794.0 | 483.0 | 1.0 |
| High | 51.0 | 53.0 | 377.0 |
| Low | 508.0 | 954.0 | 1.0 |
| [total] | 2353.0 | 1490.0 | 379.0 |
| **MCV** | | | |
| Normal | 1912.0 | 587.0 | 1.0 |
| High | 58.0 | 60.0 | 377.0 |
| Low | 383.0 | 843.0 | 1.0 |
| [total] | 2353.0 | 1490.0 | 379.0 |
| **RDW** | | | |
| Normal | 1831.0 | 362.0 | 1.0 |
| High | 519.0 | 1100.0 | 377.0 |
| Low | 3.0 | 28.0 | 1.0 |
| [total] | 2353.0 | 1490.0 | 379.0 |

**Fig. 3.** Resulted values support JRip rule classifier

As the Naive Bayes Classifier with data split of 90% gave 100% accuracy, we further investigate the produced result. The resulted values (see Fig. 3) support JRip rule classifier where it proved that HGB, MCH, MCV and RDW contribute the most in Macrocytic Anemia class. In contrast, B12 doesn't affect the appearance of Macrocytic Anemia as we assume in the beginning of this study. This was concluded from being most of the patients with Macrocytic Anemia have normal range of B12. We can derive from that, is the Macrocytic Anemia could be caused or affected by other reasons.

## 5   Conclusion

Vitamin B12 deficiencies have many effects on various body systems more profound effect is on rapidly cell including blood cell. Macrocytic anemia is one of the blood cell diseases, which is produced by reduction of the red blood cell. A brief study of the effect of vitamin B12 deficiency on Macrocytic anemia is presented in this paper. The Complete blood count (CBC) features were used for the study. Various experiments were performed based on four classification methods and shown that the JRip and Multilayer Perceptron have the higher accuracy with 100%. All experiments show that B12 doesn't affect the appearance of Macrocytic Anemia. JRip classifier proved that the attributes Hgb is the main feature to indicate the Macrocytic Anemia existence but indicator value is low.

From the medical point of view there are other causes could mask the effect of vitamin B12 deficiency on hemoglobin level and other red blood cell (RBC) variables like iron deficiency. As future work, we can study appearance Macrocytic Anemia based on other variables such as iron deficiency.

## References

1. Haykin, S.: Neural Networks: A Comprehensive Foundation, 1st edn. Prentice Hall PTR, Upper Saddle River (1994)
2. Cichosz, P.: Naïve Bayes classifier. In: Data Mining Algorithms, pp. 118–133. Wiley (2015)
3. Patil, T., Sherekar, S.: Performance analysis of Naive Bayes and J48 classification algorithm for data classification. TechRepublic. https://www.techrepublic.com/resource-library/whitepapers/performance-analysis-of-naive-bayes-and-j48-classification-algorithm-for-data-classification/. Accessed 07 Dec 2017
4. Rajput, A., Aharwal, R., Dubey, M., Raghuvanshi, M.: J48 and JRIP rules for E-governance data. http://www.cscjournals.org/library/manuscriptinfo.php?mc=IJCSS-448. Accessed 07 Dec 2017
5. Saichanma, S., Chulsomlee, S., Thangrua, N., Pongsuchart, P., Sanmun, D.: The observation report of red blood cell morphology in thailand teenager by using data mining technique. In: Advances in Hematology (2014)
6. Abdullah, M., Al-Asmari, S.: Anemia types prediction based on data mining classification algorithms, November 2016
7. Sanap, S.A., Nagori, M., Kshirsagar, V.: Classification of anemia using data mining techniques. In: Swarm, Evolutionary, and Memetic Computing, pp. 113–121 (2011)
8. Vijayarani, S., Sudha, S.: An efficient clustering algorithm for predicting diseases from hemogram blood test samples. Indian J. Sci. Technol. 8(17) (2015)
9. Jain, A.K.: Data clustering: 50 years beyond K-means. Pattern Recognit. Lett. 31(8), 651–666 (2010)
10. Ferraz, A., Brito, J.H., Carvalho, V., Machado, J.: Blood type classification using computer vision and machine learning. Neural Comput. Appl. 28(8), 2029–2040 (2017)

11. Alshami, I.: Automated diagnosis of thalassemia based on datamining classifiers. In: International Conference Information Application, ICIA 2012, pp. 440–445 (2012)
12. Dinakaran, K., Preethi, R.: A novel approach to uncover the patient blood related diseases using data mining techniques. J. Med. Sci. **13**(2), 95–102 (2013)
13. Minnie, D., Srinivasan, S.: Clustering the preprocessed automated blood cell counter data using modified K-means algorithms and generation of association rules. Int. J. Comput. Appl. **52**, 38–42 (2012)
14. El-Halees, A., Shurrab, A.: Blood tumor prediction using data mining techniques. In: Hematology Diseases, Blood Tumor, Rule Induction, Association Rules, Deep Learning, vol. 6, pp. 23–30 (2017)
15. Matsuoka, K., Yokoyama, S., Watanabe, K., Tsumoto, S.: Data mining analysis of relationship between blood stream infection and clinical background in patients undergoing lactobacillus therapy, vol. 6, pp. 1940–1945. Springer (2007)
16. Soley-Bori, M.: Dealing with missing data: key assumptions and methods for applied analysis. School of Public Health Department of Health Policy & Management, Boston University, Technical report 4, May 2013
17. Chawla, N.V., Bowyer, K.W., Hall, L.O., Kegelmeyer, W.P.: SMOTE: synthetic minority over-sampling technique. ArXiv11061813 Cs, June 2011
18. Neukart, F., Grigorescu, C.M., Moraru, S.A.: High order computational intelligence in data mining a generic approach to systemic intelligent data mining. In: 2011 6th Conference on Speech Technology and Human-Computer Dialogue (SpeD), pp. 1–9 (2011)
19. Dimitoglou, G., Adams, J.A., Jim, C.M.: Comparison of the C4.5 and a Naive Bayes classifier for the prediction of lung cancer survivability. ArXiv12061121 Cs, June 2012

# Author Index

Printed in the United States
By Bookmasters